Crop Production
in
Stress Environment

Crop Production in Stress Environment

Charlotte Jhonson

Oxford Book Company
Jaipur ▪ New Delhi

Published By

Oxford Book Company

Regd. Off. "Bony Residency", Gate No. 2
Opp. Tilak Public School, Vishweswariya Nagar,
Gopalpura Road, Jaipur - 302018, India
Ph: 0141 - 2761280, Telefax: 2761381
email - oxfordbookjpr@gmail.com
website - oxfordbookcompany.com

Delhi Office: 102, 1st Floor, Satyam House,
4327/3, Ansari Road, Darya Ganj,
New Delhi - 110002, India
Ph: 011-45652440

Published in India, 2025

ISBN : 9789380179506

Price : ₹ 2195/-

Typeset by Shivangi Computers, Jaipur

Printed and bound in India by Thomson Press Ltd., New Delhi

Preface

Crop production systems are extremely meaningful in the context of dwindling natural resources and rising population pressures. Abiotic stresses are the principle sources of yield losses. Sustaining and improving crop production in abiotic stress environments is feasible with new technologies and scientific understanding.

In order to meet the food, feed, fiber, fuel needs of ever increasing human and animal population world over, particularly in developing nations a commensurate enhancement in crop production has to be in place. This would need a holistic system approach to harness natural resources for the welfare of mankind through scientific management of land and water resources, use of appropriate agro-technology and judicious exploitation of biodiversity and genetic options for enhancing crop production especially in stress environments.

Moreover, climatic changes such as increased atmospheric temperature, uncertainty in rainfall; perpetual droughts and floods of varying intensities etc lead to environmental fragility which adversely affects crop production. Environmental concerns in terms of depletion of natural resource base including soil degradation due to over mining of nutrients, imbalanced fertilization, salinity due to excessive irrigation, water deficits and environmental pollution affecting food chain have already become apparent and points to ponder.

The book presents the much needed information on various hostile factors like stresses of moisture, temperature, salinity and acidity and their efficient management. This book covers information for students and researchers on molecular genetic approaches.

Charlotte Jhonson

Contents

Chapter 1

Introduction to Evapotranspiration

This chapter explains the concepts of and the differences between reference crop evapotranspiration (ET_o) and crop evapotranspiration under standard conditions (ET_c) and various management and environmental conditions ($ET_{c\ adj}$). It also examines the factors that affect evapotranspiration, the units in which it is normally expressed and the way in which it can be determined.

EVAPOTRANSPIRATION PROCESS

The combination of two separate processes whereby water is lost on the one hand from the soil surface by evaporation and on the other hand from the crop by transpiration is referred to as evapotranspiration (ET).

Evaporation

Evaporation is the process whereby liquid water is converted to water vapour (vaporization) and removed from the evaporating surface (vapour removal). Water evaporates from a variety of surfaces, such as lakes, rivers, pavements, soils and wet vegetation.

Energy is required to change the state of the molecules of water from liquid to vapour. Direct solar radiation and, to a lesser extent, the ambient temperature of the air provide this energy. The driving force to remove water vapour from the evaporating surface is the difference between the water vapour pressure at the evaporating surface and that of the surrounding atmosphere.

As evaporation proceeds, the surrounding air becomes gradually saturated and the process will slow down and might stop if the wet air is not transferred to the atmosphere. The replacement of the saturated air with drier air depends greatly on wind speed. Hence, solar radiation, air temperature, air humidity and wind speed are climatological parameters to consider when assessing the evaporation process.

Where the evaporating surface is the soil surface, the degree of shading of the crop canopy and the amount of water available at the evaporating surface are other factors that affect the evaporation process. Frequent rains, irrigation and water transported upwards in a soil from a shallow water Table wet the soil surface. Where the soil is able to supply water fast enough to satisfy the evaporation demand, the evaporation from the soil is determined only by the meteorological conditions.

However, where the interval between rains and irrigation becomes large and the ability of the soil to conduct moisture to pear the surface is small, the water content in the topsoil drops and the soil surface dries out. Under these circumstances the limited availability of water exerts a controlling influence on soil evaporation. In the absence of any supply of water to the soil surface, evaporation decreases rapidly and may cease almost completely within a few days.

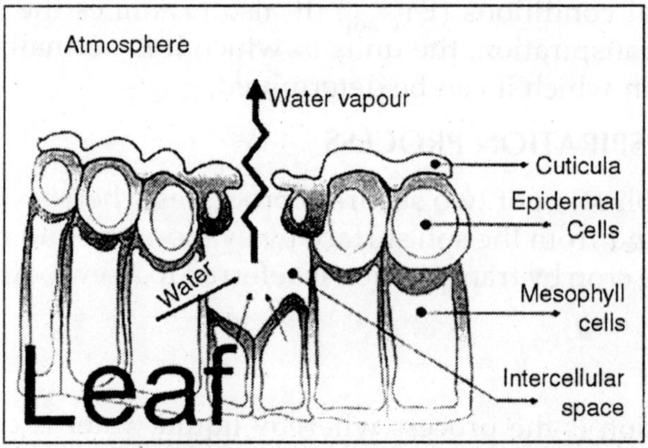

Fig. Schematic Representation of a Stoma

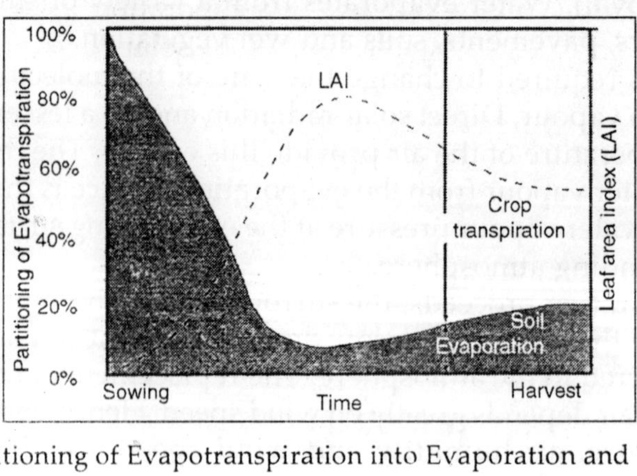

Fig. The Partitioning of Evapotranspiration into Evaporation and Transpiration over the Growing Period for an Annual Field Crop

Transpiration

Transpiration consists of the vaporization of liquid water contained in plant tissues and the vapour removal to the atmosphere. Crops predominately lose their water through stomata. These are small openings on the plant leaf through which gases and water vapour pass. The water, together with some nutrients, is taken up by the roots and transported through the plant.

The vaporization occurs within the leaf, namely in the intercellular spaces, and the vapour exchange with the atmosphere is controlled by the stomatal aperture.

Nearly all water taken up is lost by transpiration and only a tiny fraction is used within the plant.

Transpiration, like direct evaporation, depends on the energy supply, vapour pressure gradient and wind. Hence, radiation, air temperature, air humidity and wind terms should be considered when assessing transpiration.

The soil water content and the ability of the soil to conduct water to the roots also determine the transpiration rate, as do waterlogging and soil water salinity.

The transpiration rate is also influenced by crop characteristics, environmental aspects and cultivation practices. Different kinds of plants may have different transpiration rates. Not only the type of crop, but also the crop development, environment and management should be considered when assessing transpiration.

EVAPOTRANSPIRATION (ET)

Evaporation and transpiration occur simultaneously and there is no easy way of distinguishing between the two processes. Apart from the water availability in the topsoil, the evaporation from a cropped soil is mainly determined by the fraction of the solar radiation reaching the soil surface.

This fraction decreases over the growing period as the crop develops and the crop canopy shades more and more of the ground area. When the crop is small, water is predominately lost by soil evaporation, but once the crop is well developed and completely covers the soil, transpiration becomes the main process.

The partitioning of evapotranspiration into evaporation and transpiration is plotted in correspondence to leaf area per unit surface of soil below it. At sowing nearly 100% of ET comes from evaporation, while at full crop cover more than 90% of ET comes from transpiration.

Units

The evapotranspiration rate is normally expressed in millimetres (mm) per unit time. The rate expresses the amount of water lost from a cropped surface in units of water depth. The time unit can be an hour, day, decade, month or even an entire growing period or year.

As one hectare has a surface of 10000 m^2 and 1 mm is equal to 0.001 m, a loss of 1 mm of water corresponds to a loss of 10 m^3 of water per hectare. In other words, 1 mm day^{-1} is equivalent to 10 m^3 ha^{-1} day^{-1}.

Water depths can also be expressed in terms of energy received per unit area. The energy refers to the energy or heat required to vaporize free water. This energy, known as the latent heat of vaporization , is a function of the water temperature. For example, at 20°C, l is about 2.45 MJ kg^{-1}.

In other words, 2.45 MJ are needed to vaporize 1 kg or 0.001 m^3 of water. Hence, an energy input of 2.45 MJ per m^2 is able to vaporize 0.001 m or 1 mm of water, and therefore 1 mm of water is equivalent to 2.45 MJ m^{-2}. The evapotranspiration rate expressed in units of MJ m^{-2} day^{-1} is represented by l ET, the latent heat flux.

Table Summarizes the units used to express the evapotranspiration rate and the conversion factors.

Table:Conversion Factors for Evapotranspiration

	depth	volume per unit area		energy per unit area *
	mm day^{-1}	m^3 ha^{-1} day^{-1}	l s^{-1} ha^{-1}	MJ m^{-2} day^{-1}
1 mm day^{-1}	1	10	0.116	2.45
1 m^3 ha^{-1} day^{-1}	0.1	1	0.012	0.245
1 l s^{-1} ha^{-1}	8.640	86.40	1	21.17
1 MJ m^{-2} day^{-1}	0.408	4.082	0.047	1

For water with a density of 1000 kg m^{-3} and at 20°C.

Example. Converting Evaporation from one Unit to Another

On a summer day, net solar energy received at a lake reaches 15 MJ per square metre per day. If 80% of the energy is used to vaporize water, how large could the depth of evaporation be?			
From Table 1:	1 MJ m^{-2} day^{-1} =	0.408	mm day^{-1}
Therefore:	0.8 x 15 MJ m^{-2} day^{-1} = 0.8 x 15 x 0.408 mm d^{-1} =	4.9	mm day^{-1}
The evaporation rate could be 4.9 mm/day			

Factors Affecting Evapotranspiration

Weather parameters, crop characteristics, management and environmental aspects are factors affecting evaporation and transpiration. The related ET concepts presented are discussed in the section on evapotranspiration concepts.

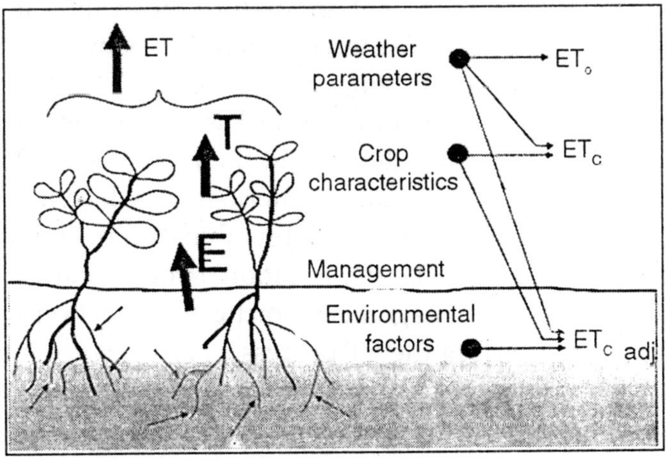

Fig. Factors Affecting Evapotranspiration with Reference to Related ET Concepts

Weather Parameters

The principal weather parameters affecting evapotranspiration are radiation, air temperature, humidity and wind speed. Several procedures have been developed to assess the evaporation rate from these parameters. The evaporation power of the atmosphere is expressed by the reference crop evapotranspiration (ET_o). The reference crop evapotranspiration represents the evapotranspiration from a standardized vegetated surface.

CROP FACTORS

The crop type, variety and development stage should be considered when assessing the evapotranspiration from crops grown in large, well-managed fields. Differences in resistance to transpiration, crop height, crop roughness, reflection, ground cover and crop rooting characteristics result in different ET levels in different types of crops under identical environmental conditions. Crop evapotranspiration under standard conditions (ET_c) refers to the evaporating demand from crops that are grown in large fields under optimum soil water, excellent management and environmental conditions, and achieve full production under the given climatic conditions.

Management and Environmental Conditions

Factors such as soil salinity, poor land fertility, limited application of fertilizers, the presence of hard or impenetrable soil horizons, the absence of control of diseases and pests and poor soil management may limit the crop development and reduce the evapotranspiration.

Other factors to be considered when assessing ET are ground cover, plant density and the soil water content.

The effect of soil water content on ET is conditioned primarily by the magnitude of the water deficit and the type of soil. On the other hand, too much water will result in waterlogging which might damage the root and limit root water uptake by inhibiting respiration.

When assessing the ET rate, additional consideration should be given to the range of management practices that act on the climatic and crop factors affecting the ET process.

Cultivation practices and the type of irrigation method can alter the microclimate, affect the crop characteristics or affect the wetting of the soil and crop surface.

A windbreak reduces wind velocities and decreases the ET rate of the field directly beyond the barrier.

The effect can be significant especially in windy, warm and dry conditions although evapotranspiration from the trees themselves may offset any reduction in the field.

Soil evaporation in a young orchard, where trees are widely spaced, can be reduced by using a well-designed drip or trickle irrigation system.

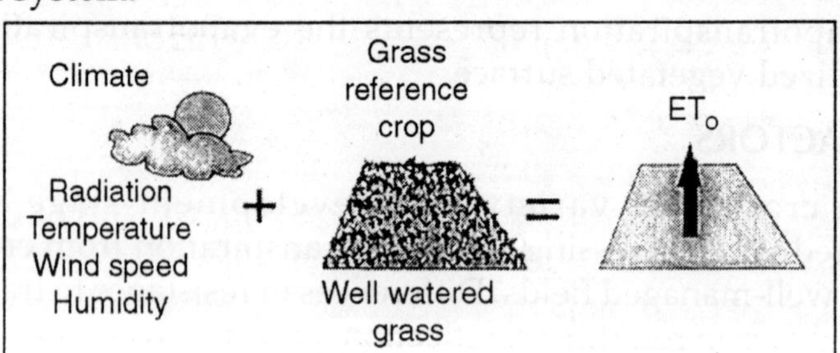

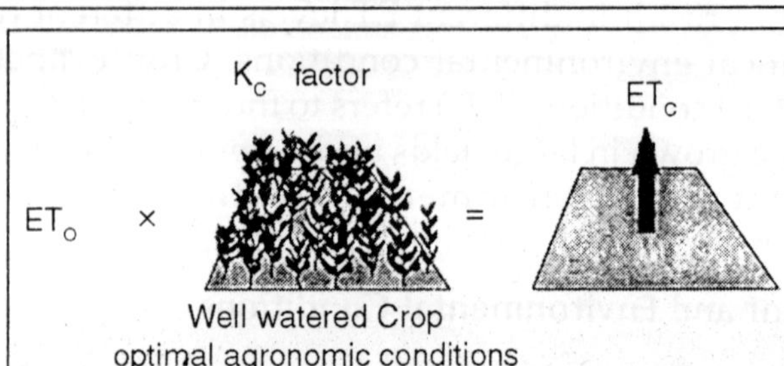

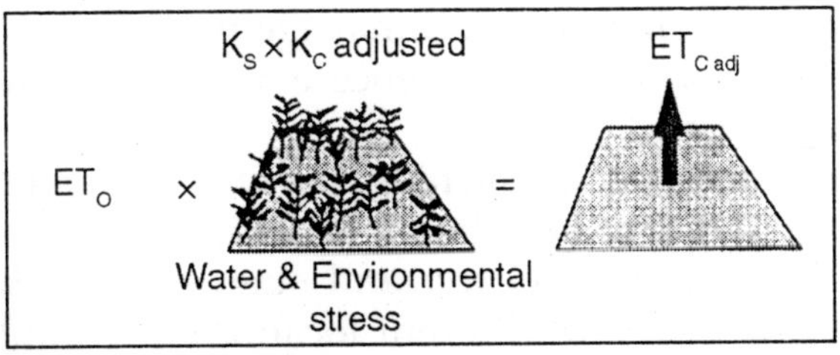

Fig. Reference (ET_o), Crop Evapotranspiration under Standard (ET_c) and Non-Standard Conditions ($ET_{c\ adj}$)

The drippers apply water directly to the soil near trees, thereby leaving the major part of the soil surface dry, and limiting the evaporation losses. The use of mulches, especially when the crop is small, is another way of substantially reducing soil evaporation. Anti-transpirants, such as stomata-closing, film-forming or reflecting material, reduce the water losses from the crop and hence the transpiration rate.

Where field conditions differ from the standard conditions, correction factors are required to adjust ET_c. The adjustment reflects the effect on crop evapotranspiration of the environmental and management conditions in the field.

Evapotranspiration Concepts

Distinctions are made between reference crop evapotranspiration (ET_o), crop evapotranspiration under standard conditions (ET_c) and crop evapotranspiration under non-standard conditions ($ET_{c\ adj}$). ET_o is a climatic parameter expressing the evaporation power of the atmosphere. ET_c refers to the evapotranspiration from excellently managed, large, well-watered fields that achieve full production under the given climatic conditions.

Due to sub optimal crop management and environmental constraints that affect crop growth and limit evapotranspiration, ET_c under non-standard conditions generally requires a correction.

REFERENCE CROP EVAPOTRANSPIRATION (ET_o)

The evapotranspiration rate from a reference surface, not short of water, is called the reference crop evapotranspiration or reference evapotranspiration and is denoted as ET_o. The reference surface is a hypothetical grass reference crop with specific characteristics. The use of other denominations such as potential ET is strongly discouraged due to ambiguities in their definitions.

The concept of the reference evapotranspiration was introduced to study the evaporative demand of the atmosphere independently of crop type, crop development and management practices. As water is abundantly available at the reference evapotranspiring surface, soil factors do not affect ET.

Relating ET to a specific surface provides a reference to which ET from other surfaces can be related. It obviates the need to define a separate ET level for each crop and stage of growth. ET_o values measured or calculated at different locations or in different seasons are comparable as they refer to the ET from the same reference surface.

The only factors affecting ET_o are climatic parameters. Consequently, ET_o is a climatic parameter and can be computed from weather data. ET_o expresses the evaporating power of the atmosphere at a specific location and time of the year and does not consider the crop characteristics and soil factors.

The FAO Penman-Monteith method is recommended as the sole method for determining ET_o. The method has been selected because it closely approximates grass ET_o at the location evaluated, is physically based, and explicitly incorporates both physiological and aerodynamic parameters. Moreover, procedures have been developed for estimating missing climatic parameters.

Typical ranges for ET_o values for different agroclimatic regions. These values are intended to familiarize inexperienced users with typical ranges, and are not intended for direct application.

Crop Evapotranspiration under Standard Conditions (ET_c)

The crop evapotranspiration under standard conditions, denoted as ET_c, is the evapotranspiration from disease-free, well-fertilized crops, grown in large fields, under optimum soil water conditions, and achieving full production under the given climatic conditions.

Table. Average ET_o for Different Agroclimatic Regions in mm/Day.

Regions	Mean daily temperature (°C)		
	Cool ~10°C	Moderate 20°C	Warm > 30°C
Tropics and subtropics			
- humid and sub-humid	2 - 3	3 - 5	5 - 7
-arid and semi-arid	2 - 4	4 - 6	6 - 8
Temperate region			
- humid and sub-humid	1 - 2	2 - 4	4 - 7
-arid and semi-arid	1 - 3	4 - 7	6 - 9

The amount of water required to compensate the evapotranspiration loss from the cropped field is defined as crop water requirement. Although the values for crop evapotranspiration and crop water requirement are identical, crop water requirement refers to the amount of water that needs to be supplied, while crop evapotranspiration refers to the amount of water that is lost through evapotranspiration. The irrigation water requirement basically represents the difference between the crop water requirement and effective precipitation. The irrigation water requirement also includes additional water for leaching of salts and to compensate for non-uniformity of water application. Calculation of the irrigation water requirement is not covered in this publication, but will be the topic of a future Irrigation and Drainage Paper.

Crop evapotranspiration can be calculated from climatic data and by integrating directly the crop resistance, albedo and air resistance factors in the Penman-Monteith approach. As there is still a considerable lack of information for different crops, the Penman-Monteith method is used for the estimation of the standard reference crop to determine its evapotranspiration rate, i.e., ET_o. Experimentally determined ratios of ET_c/ET_o, called crop coefficients (K_c), are used to relate ET_c to

$$ET_o \text{ or } ET_c = K_c\, ET_o.$$

Differences in leaf anatomy, stomatal characteristics, aerodynamic properties and even albedo cause the crop evapotranspiration to differ from the reference crop evapotranspiration under the same climatic conditions. Due to variations in the crop characteristics throughout its growing season, K_c for a given crop changes from sowing till harvest. The calculation of crop evapotranspiration under standard conditions (ET_c).

Crop Evapotranspiration under Non-standard Conditions ($ET_{c\ adj}$)

The crop evapotranspiration under non-standard conditions ($ET_{c\ adj}$) is the evapotranspiration from crops grown under management and environmental conditions that differ from the standard conditions. When cultivating crops in fields, the real crop evapotranspiration may deviate from ET_c due to non-optimal conditions such as the presence of pests and diseases, soil salinity, low soil fertility, water shortage or waterlogging. This may result in scanty plant growth, low plant density and may reduce the evapotranspiration rate below ET_c.

The crop evapotranspiration under non-standard conditions is calculated by using a water stress coefficient K_s and/or by adjusting K_c for all kinds of other stresses and environmental constraints on crop evapotranspiration. The adjustment to ET_c for water stress, management and environmental constraints is discussed in Part C of this handbook.

DETERMINING EVAPOTRANSPIRATION

ET Measurement

Evapotranspiration is not easy to measure. Specific devices and accurate measurements of various physical parameters or the soil water balance in lysimeters are required to determine evapotranspiration. The methods are often expensive, demanding in terms of accuracy of measurement and can only be fully exploited by well-trained research personnel.

Although the methods are inappropriate for routine measurements, they remain important for the evaluation of ET estimates obtained by more indirect methods.

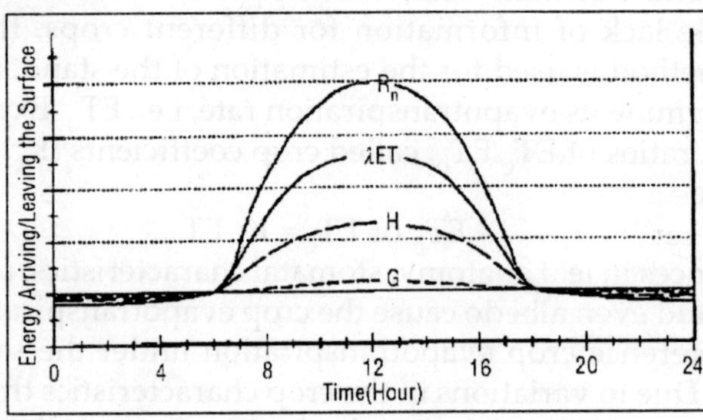

Fig. Schematic Presentation of the Diurnal Variation of the Components of the Energy Balance Above a well-watered Transpiring Surface on a Cloudless Day.

Energy Balance and Microclimatological Methods

Evaporation of water requires relatively large amounts of energy, either in the form of sensible heat or radiant energy. Therefore the evapotranspiration process is governed by energy exchange at the vegetation surface and is limited by the amount of energy available.

Because of this limitation, it is possible to predict the evapotranspiration rate by applying the principle of energy conservation. The energy arriving at the surface must equal the energy leaving the surface for the same time period.

All fluxes of energy should be considered when deriving an energy balance equation. The equation for an evaporating surface can be written as:

$$R_n - \Gamma - l\,ET - H = 0$$

where R_n is the net radiation, H the sensible heat, G the soil heat flux and l ET the latent heat flux. The various terms can be either positive or negative. Positive R_n supplies energy to the surface and positive G, l ET and H remove energy from the surface.

In Equation 1 only vertical fluxes are considered and the net rate at which energy is being transferred horizontally, by advection, is ignored. Therefore the equation is to be applied to large, extensive surfaces of homogeneous vegetation only.

The equation is restricted to the four components: R_n, l ET, H and G. Other energy terms, such as heat stored or released in the plant, or the energy used in metabolic activities, are not considered.These terms account for only a small fraction of the daily net radiation and can be considered negligible when compared with the other four components.

The latent heat flux (l ET) representing the evapotranspiration fraction can be derived from the energy balance equation if all other components are known. Net radiation (R_n) and soil heat fluxes (G) can be measured or estimated from climatic parameters. Measurements of the sensible heat (H) are however complex and cannot be easily obtained. H requires accurate measurement of temperature gradients above the surface.

Another method of estimating evapotranspiration is the mass transfer method. This approach considers the vertical movement of small parcels of air (eddies) above a large homogeneous surface. The eddies transport material (water vapour) and energy (heat, momentum) from and towards the evaporating surface. By assuming steady state conditions and that the eddy transfer coefficients for water vapour are proportional to those for heat and momentum, the evapotranspiration rate can be computed from the vertical gradients of air temperature and water vapour via the Bowen ratio.

Other direct measurement methods use gradients of wind speed and water vapour. These methods and other methods such as eddy covariance, require accurate measurement of vapour pressure, and air temperature or wind speed at different levels above the surface. Therefore, their application is restricted to primarily research situations.

Soil Water Balance

Evapotranspiration can also be determined by measuring the various components of the soil water balance. The method consists of assessing the incoming and outgoing water flux into the crop root zone over some time period. Irrigation (I) and rainfall (P) add water to the root zone. Part of I and P might be lost by surface runoff (RO) and by deep percolation (DP) that will eventually recharge the water Table.

Water might also be transported upward by capillary rise (CR) from a shallow water Table towards the root zone or even transferred horizontally by subsurface flow in (SF_{in}) or out of (SF_{out}) the root zone. In many situations, however, except under conditions with large slopes, SF_{in} and SF_{out} are minor and can be ignored. Soil evaporation and crop transpiration deplete water from the root zone. If all fluxes other than evapotranspiration (ET) can be assessed, the evapotranspiration can be deduced from the change in soil water content (D SW) over the time period:

$$ET = I + P - RO - DP + CR \pm D\,SF \pm D\,SW$$

Some fluxes such as subsurface flow, deep percolation and capillary rise from a water Table are difficult to assess and short time periods cannot be considered. The soil water balance method can usually only give ET estimates over long time periods of the order of week-long or ten-day periods.

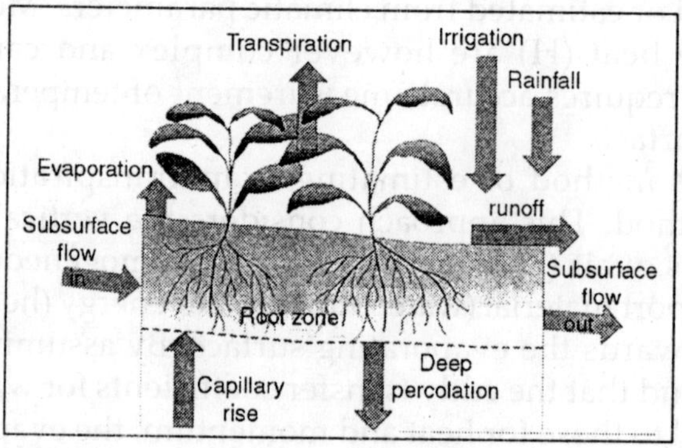

Fig. Soil Water Balance of the Root Zone

Lysimeters

By isolating the crop root zone from its environment and controlling the processes that are difficult to measure, the different terms in the soil water balance equation can be determined with greater accuracy. This is done in lysimeters where the crop grows in isolated tanks filled with either disturbed or undisturbed soil.

In precision weighing lysimeters, where the water loss is directly measured by the change of mass, evapotranspiration can be obtained with an accuracy of a few hundredths of a millimetre, and small time periods such as an hour can be considered. In non-weighing lysimeters the evapotranspiration for a given time period is determined by deducting the drainage water, collected at the bottom of the lysimeters, from the total water input.

A requirement of lysimeters is that the vegetation both inside and immediately outside of the lysimeter be perfectly matched (same height and leaf area index). This requirement has historically not been closely adhered to in a majority of lysimeter studies and has resulted in severely erroneous and unrepresentative ET_c and K_c data.

As lysimeters are difficult and expensive to construct and as their operation and maintenance require special care, their use is limited to specific research purposes.

ET COMPUTED FROM METEOROLOGICAL DATA

Owing to the difficulty of obtaining accurate field measurements, ET is commonly computed from weather data. A large number of empirical or semi-empirical equations have been developed for assessing crop or reference crop evapotranspiration from meteorological data. Some of the methods are only valid under specific climatic and agronomic conditions and cannot be applied under conditions different from those under which they were originally developed.

Numerous researchers have analysed the performance of the various calculation methods for different locations. As a result of an Expert Consultation held in May 1990, the FAO Penman-Monteith method is now recommended as the standard method for the definition and computation of the reference evapotranspiration, ET_o.

The ET from crop surfaces under standard conditions is determined by crop coefficients (K_c) that relate ET_c to ET_o. The ET from crop surfaces under non-standard conditions is adjusted by a water stress coefficient (K_s) and/or by modifying the crop coefficient.

ET ESTIMATED FROM PAN EVAPORATION

Evaporation from an open water surface provides an index of the integrated effect of radiation, air temperature, air humidity and wind on evapotranspiration. However, differences in the water and cropped surface produce significant differences in the water loss from an open water surface and the crop.

The pan has proved its practical value and has been used successfully to estimate reference evapotranspiration by observing the evaporation loss from a water surface and applying empirical coefficients to relate pan evaporation to ET_o.

CROP EVAPOTRANSPIRATION (ET_c)

This chapter outlines the crop coefficient approach for calculating the crop evapotranspiration under standard conditions (ET_c). The standard conditions refer to crops grown in large fields under excellent agronomic and soil water conditions. The crop evapotranspiration differs distinctly from the reference evapotranspiration (ET_o) as the ground cover, canopy properties and aerodynamic resistance of the crop are different from grass.

The effects of characteristics that distinguish field crops from grass are integrated into the crop coefficient (K_c). In the crop coefficient approach, crop evapotranspiration is calculated by multiplying ET_o by K_c.

Differences in evaporation and transpiration between field crops and the reference grass surface can be integrated in a single crop coefficient (K_c) or separated into two coefficients: a basal crop (K_{cb}) and a soil evaporation coefficient (K_e), i.e., $K_c = K_{cb} + K_e$. The approach to follow should be selected as a function of the purpose of the calculation, the accuracy required and the data available.

CALCULATION PROCEDURES

Direct Calculation

The evapotranspiration rate from a cropped surface can be directly measured by the mass transfer or the energy balance method. It can also be derived from studies of the soil water balance determined from cropped fields or from lysimeters.

Crop evapotranspiration can also be derived from meteorological and crop data by means of the Penman-Monteith equation. By adjusting the albedo and the aerodynamic and canopy surface resistances to the growing characteristics of the specific crop, the evapotranspiration rate can be directly estimated. The albedo and resistances are, however, difficult to estimate accurately as they may vary continually during the growing season as climatic conditions change, as the crop develops, and with wetness of the soil surface. The canopy resistance will further be influenced by the soil water availability, and it increases strongly if the crop is subjected to water stress.

As there is still a considerable lack of consolidated information on the aerodynamic and canopy resistances for the various cropped surfaces, the FAO Penman-Monteith method is used in this handbook only for estimating ET_o, the evapotranspiration from a well-watered hypothetical grass surface having fixed crop height, albedo and surface resistance.

Crop Coefficient Approach

In the crop coefficient approach the crop evapotranspiration, ET_c, is calculated by multiplying the reference crop evapotranspiration, ET_o, by a crop coefficient, K_c:

$$ET_c = K_c\, ET_o$$

where

ET_c crop evapotranspiration [mm d^{-1}],

K_c crop coefficient [dimensionless],

ET_o reference crop evapotranspiration [mm d^{-1}].

Most of the effects of the various weather conditions are incorporated into the ET_o estimate. Therefore, as ET_o represents an index of climatic demand, K_c varies predominately with the specific crop characteristics and only to a limited extent with climate. This enables the transfer of standard values for K_c between locations and between climates. This has been a primary reason for the global acceptance and usefulness of the crop coefficient approach and the K_c factors developed in past studies.

The reference ET_o is defined and calculated using the FAO Penman-Monteith equation. The crop coefficient, K_c, is basically the ratio of the crop ET_c to the reference ET_o, and it represents an integration of the effects of four primary characteristics that distinguish the crop from reference grass. These characteristics are:

- Crop height. The crop height influences the aerodynamic resistance term, r_a, of the FAO Penman-Monteith equation and the turbulent transfer of vapour from the crop into the atmosphere. The r_a tenu appears twice in me full form of the FAO Penman-Monteith equation.
- Albedo (reflectance) of the crop-soil surface. The albedo is affected by the fraction of ground covered by vegetation and by the soil surface wetness. The albedo of the crop-soil surface influences the net radiation of the surface, R_n, which is the primary source of the energy exchange for the evaporation process.
- Canopy resistance. The resistance of me crop to vapour transfer is affected by leaf area (number of stomata), leaf age and condition, and the degree of stomatal control. The canopy resistance influences the surface resistance, r_s.
- Evaporation from soil, especially exposed soil.

The soil surface wetness and the fraction of ground covered by vegetation influence the surface resistance, r_s. Following soil wetting, the vapour transfer rate from the soil is high, especially for crops having incomplete ground cover. The combined surface resistance of the canopy and of the soil determines the (bulk) surface resistance, r_s. The surface resistance term in the Penman-Monteith equation represents the resistance to vapour flow from within plant leaves and from beneath the soil surface.

The K_c in Equation 56 predicts ET_c under standard conditions. This represents the upper envelope of crop evapotranspiration and represents conditions where no limitations are placed on crop growth or evapotranspiration due to water shortage, crop density, or disease, weed, insect or salinity pressures.

The ET_c predicted by K_c is adjusted if necessary to non-standard conditions, $ET_{c\ adj}$, where any environmental condition or characteristic is known to have an impact on or to limit ET_c. Factors for correcting ET_c to $ET_{c\ adj}$ are described in Part C.

Factors Determining the Crop Coefficient

The crop coefficient integrates the effect of characteristics that distinguish a typical field crop from the grass reference, which has a constant appearance and a complete ground cover. Consequently, different crops will have different K_c coefficients.

The changing characteristics of the crop over the growing season also affect the K_c coefficient. Finally, as evaporation is an integrated part of crop evapotranspiration, conditions affecting soil evaporation will also have an effect on K_c.

Crop Type

Due to differences in albedo, crop height, aerodynamic properties, and leaf and stomata properties, the evapotranspiration from full grown, well-watered crops differs from ET_o.

The close spacings of plants and taller canopy height and roughness of many full grown agricultural crops cause these crops to have K_c factors that are larger than 1. The K_c factor is often 5-10% higher than the reference (where $K_c = 1.0$), and even 15-20% greater for some tall crops such as maize, sorghum or sugar cane. Typical values for the crop coefficient for full grown crops ($K_{c\ mid}$).

Crops such as pineapples, that close their stomata during the day, have very small crop coefficients. In most species, however, the stomata open as irradiance increases. In addition to the stomatal response to environment, the position and number of the stomata and the resistance of the cuticula to vapour transfer determine the water loss from the crop.

Species with stomata on only the lower side of the leaf and/or large leaf resistances will have relatively smaller K_c values. This is the case for citrus and most deciduous fruit trees. Transpiration control and spacing of the trees, providing only 70% ground cover for mature trees, may cause the K_c of those trees, if cultivated without a ground cover crop, to be smaller than on).

Climate

The K_c values of Table are typical values expected for average K_c under a standard climatic condition, which is defined as a sub-humid climate with average daytime minimum relative humidity (RH_{min}) ≈ 45% and having calm to moderate wind speeds averaging 2 m/s.

Variations in wind alter the aerodynamic resistance of the crops and hence their crop coefficients, especially for those crops that are substantially taller than the hypothetical grass reference. The effect of the difference in aerodynamic properties between the grass reference surface and agricultural crops is not only crop specific. It also varies with the climatic conditions and crop height.

Because aerodynamic properties are greater for many agricultural crops as compared to the grass reference, the ratio of ET_c to ET_o (i.e., K_c) for many crops increases as wind speed increases and as relative humidity decreases. More arid climates and conditions of greater wind speed will have higher values for K_c. More humid climates and conditions of lower wind speed will have lower values for K_c.

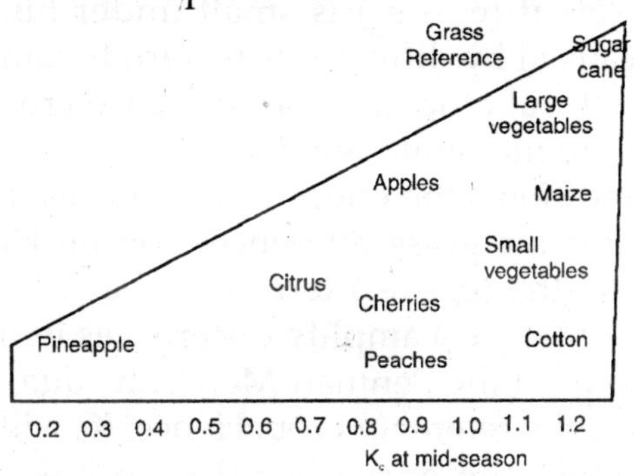

Fig. Typical K_c for Different Types of Full Grown Crops

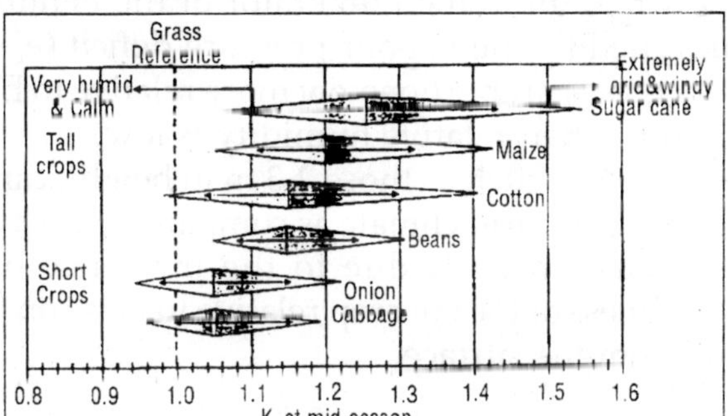

Fig. Extreme Ranges Expected in K_c for Full Grown Crops as Climate and Weather Change

The relative impact of the climate on K_c for full grown crops. The upper bounds represent extremely arid and windy conditions, while the lower bounds are valid under very humid and calm weather conditions. The ranges expected in K_c as climate and weather conditions change are quite small for short crops but are large for tall crops. Guidelines for the adjustment of K_c to the climatic conditions as a function of crop height.

Under humid and calm wind conditions, K_c becomes less dependent on the differences between the aerodynamic components of ET_c and ET_o and the K_c values for 'full-cover' agricultural crops do not exceed 1.0 by more than about 0.05. This is because full-cover agricultural crops and the reference crop of clipped grass both provide for nearly maximum absorption of shortwave radiation, which is the primary energy source for evaporation under humid and calm conditions.

Generally, the albedos, a, are similar over a wide range of full-cover agricultural crops, including the reference crop. Because the vapour pressure deficit $(e_s - e_a)$ is small under humid conditions, differences in ET caused by differences in aerodynamic resistance, r_a, between the agricultural crop and the reference crop are also small, especially with low to moderate wind speed.

Under arid conditions, the effect of differences in r_a between the agricultural crop and the grass reference crop on ET_c become more pronounced because the $(e_s - e_a)$ term may be relatively large. The larger magnitudes of $(e_s - e_a)$ amplify differences in the aerodynamic term in the numerator of the Penman-Monteith equation (Equation 3) for both the crop and the reference crop. Hence, K_c will be larger under arid conditions when the agricultural crop has a leaf area and roughness height that are greater than that of the grass reference.

Because the $1/r_a$ term in the numerator of the Penman-Monteith equation is multiplied by the vapour pressure deficit $(e_s - e_a)$, the ET from tall crops increases proportionately more relative to ET_o than does ET from short crops when relative humidity is low.

The K_c for tall crops, such as those 2-3 m in height, can be as much as 30% higher in a windy, arid climate as compared with a calm, humid climate. The increase in K_c is due to the influence of the larger aerodynamic roughness of the tall crop relative to grass on the transport of water vapour from the surface.

Soil Evaporation

Differences in soil evaporation and crop transpiration between field crops and the reference surface are integrated within the crop coefficient. The K_c coefficient for full-cover crops primarily reflects differences in transpiration as the contribution of soil evaporation is relatively small.

After rainfall or irrigation, the effect of evaporation is predominant when the crop is small and scarcely shades the ground. For such low-cover conditions, the K_c coefficient is determined largely by the frequency with which the soil surface is wetted.

Where the soil is wet for most of the time from irrigation or rain, the evaporation from the soil surface will be considerable and K_c may exceed 1. On the other hand, where the soil surface is dry, evaporation is restricted and K_c will be small and might even drop to as low as 0.1.

Differences in soil evaporation between the field crop and the reference surface can be forecast more precisely by using a dual crop coefficient.

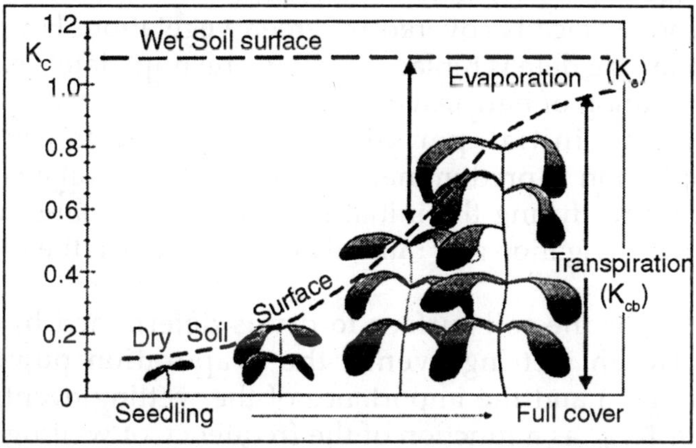

Fig. The Effect of Evaporation on K_c. the Horizontal Line Represents K_c when the Soil Surface is kept Continuously Wet

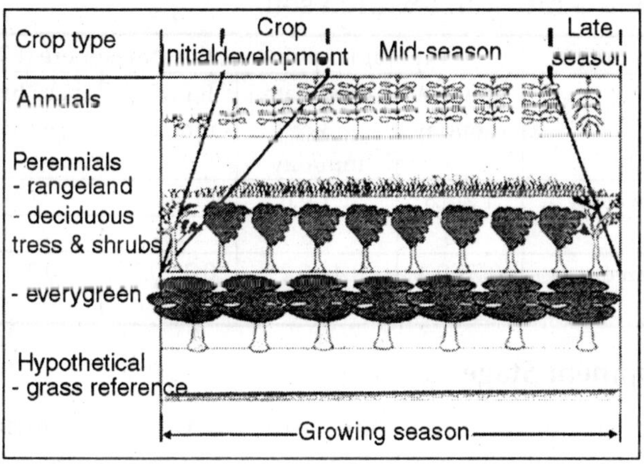

Fig. Crop Growth Stages for Different Types of Crops

Crop Growth Stages

As the crop develops, the ground cover, crop height and the leaf area change. Due to differences in evapotranspiration during the various growth stages, the K_c for a given crop will vary over the growing period. The growing period can be divided into four distinct growth stages: initial, crop development, mid-season and late season. The general sequence and proportion of these stages for different types of crops.

Initial Stage

The initial stage runs from planting date to approximately 10% ground cover. The length of the initial period is highly dependent on the crop, the crop variety, the planting date and the climate. The end of the initial period is determined as the time when approximately 10% of the ground surface is covered by green vegetation. For perennial crops, the planting date is replaced by the 'greenup' date, i.e., the time when the initiation of new leaves occurs.

During the initial period, the leaf area is small, and evapotranspiration is predominately in the form of soil evaporation. Therefore, the K_c during the initial period ($K_{c\ ini}$) is large when the soil is wet from irrigation and rainfall and is low when the soil surface is dry.

The time for the soil surface to dry is determined by the time interval between wetting events, the evaporation power of the atmosphere (ET_o) and the importance of the wetting event. General estimates for $K_{c\ ini}$ as a function of the frequency of wetting and ET_o. The data assume a medium textured soil.

Table. Approximate values for $K_{c\ ini}$ for medium wetting events (10-40 mm) and a medium textured soil.

wetting interval	evaporating power of the atmosphere (ETo)			
	low 1-3 mm/day	moderate 3-5 mm/day	high 5-7 mm/day	very high > 7 mm/day
less than weekly	1.2-0.8	1.1-0.6	1.0-0.4	0.9-0.3
weekly	0.8	0.6	0.4	0.3
longer than once per week	0.7 - 0.4	0.4 - 0.2*	0.3 - 0.2*	0.2*- 0.1*

Crop Development Stage

The crop development stage runs from 10% ground cover to effective full cover. Effective full cover for many crops occurs at the initiation of flowering. For row crops where rows commonly interlock leaves such as beans, sugar beets, potatoes and corn, effective cover can be defined as the time when some leaves of plants in adjacent rows begin to intermingle so that soil shading becomes nearly complete, or when plants reach nearly full size if no intermingling occurs.

For some crops, especially those taller than 0.5 m, the average fraction of the ground surface covered by vegetation (f_c) at the start of effective full cover is about 0.7-0.8. Fractions of sunlit and shaded soil and leaves do not change significantly with further growth of the crop beyond $f_c \approx 0.7$ to 0.8.

It is understood that the crop or plant can continue to grow in both height and leaf area after the time of effective full cover. Because it is difficult to visually determine when densely sown vegetation such as winter and spring cereals and some grasses reach effective full cover, the more easily detecTable stage of heading (flowering) is generally used for these types of crops.

For dense grasses, effective full cover may occur at about 0.10-0.15 m height. For thin stands of grass (dry rangeland), grass height may approach 0.3-0.5 m before effective full cover is reached. Densely planted forages such as alfalfa and clover reach effective full cover at about 0.3-0.4 m.

Another way to estimate the occurrence of effective full cover is when the leaf area index (LAI) reaches three. LAI is defined as the average total area of leaves (one side) per unit area of ground surface.

As the crop develops and shades more and more of the ground, evaporation becomes more restricted and transpiration gradually becomes the major process. During the crop development stage, the K_c value corresponds to amounts of ground cover and plant development.

Typically, if the soil surface is dry, $K_c - 0.5$ corresponds to about 25-40% of the ground surface covered by vegetation due to the effects of shading and due to microscale transport of sensible heat from the soil into the vegetation. A $K_c = 0.7$ often corresponds to about 40-60% ground cover. These values will vary, depending on the crop, frequency of wetting and whether the crop uses more water than the reference crop at full ground cover (e.g., depending on its canopy architecture and crop height relative to clipped grass).

Mid-Season Stage

The mid-season stage runs from effective full cover to the start of maturity. The start of maturity is often indicated by the beginning of the ageing, yellowing or senescence of leaves, leaf drop, or the browning of fruit to the degree that the crop evapotranspiration is reduced relative to the reference ET_0. The mid-season stage is the longest stage for perennials and for many annuals, but it may be relatively short for vegetable crops that are harvested fresh for their green vegetation.

At the mid-season stage the K_c reaches its maximum value. The value for K_c ($K_{c\,mid}$) is relatively constant for most growing and cultural conditions. Deviation of the $K_{c\,mid}$ from the reference value '1' is primarily due to differences in crop height and resistance between the grass reference surface and the agricultural crop and weather conditions.

Late Season Stage

The late season stage runs from the start of maturity to harvest or full senescence. The calculation for K_c and ET_c is presumed to end when the crop is harvested, dries out naturally, reaches full senescence, or experiences leaf drop.

For some perennial vegetation in frost free climates, crops may grow year round so that the date of termination may be taken as the same as the date of 'planting'.

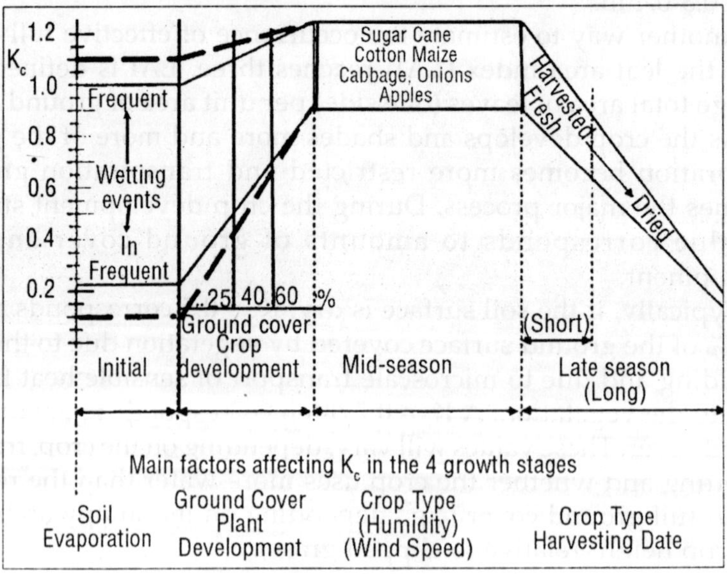

Fig. Typical Ranges Expected in K_c for the Four Growth Stages

The K_c value at the end of the late season stage ($K_{c\,end}$) reflects crop and water management practices. The $K_{c\,end}$ value is high if the crop is frequently irrigated until harvested fresh. If the crop is allowed to senesce and to dry out in the field before harvest, the $K_{c\ end}$ value will be small. Senescence is usually associated with less efficient stomatal conductance of leaf surfaces due to the effects of ageing, thereby causing a reduction in K_c.

The variation in K_c for different crops as influenced by weather factors and crop development.

Crop Evapotranspiration (ET_c)

Crop evapotranspiration is calculated by multiplying ET_0 by K_c, a coefficient expressing the difference in evapotranspiration between the cropped and reference grass surface.

The difference can be combined into one single coefficient, or it can be split into two factors describing separately the differences in evaporation and transpiration between both surfaces. The selection of the approach depends on the purpose of the calculation, the accuracy required, the climatic data available and the time step with which the calculations are executed. Table presents the general selection criteria.

Table. General selection criteria for the single and dual crop coefficient approaches

	Single crop coefficient K_c	Dual crop coefficient $K_{cb} + K_e$
Purpose of calculation	- irrigation planning and design - irrigation management - basic irrigation schedules - real time irrigation scheduling for non-frequent water applications (surface and sprinkler irrigation)	- research - real time irrigation scheduling - irrigation scheduling for high frequency water application (microirrigation and automated sprinkler irrigation) - supplemental irrigation - detailed soil and hydrologic water balance studies
Time step	daily, 10-day, monthly (data and calculation)	daily (data and calculation)
Solution method	graphical pocket calculator computer	computer

Single and Dual Crop Coefficient Approaches

Single Crop Coefficient Approach (Kc)

In the single crop coefficient approach, the effect of crop transpiration and soil evaporation are combined into a single K_c coefficient. The coefficient integrates differences in the soil evaporation and crop transpiration rate between the crop and the grass reference surface. As soil evaporation may fluctuate daily as a result of rainfall or irrigation, the single crop coefficient expresses only the time-averaged (multi-day) effects of crop evapotranspiration.

As the single K_c coefficient averages soil evaporation and transpiration, the approach is used to compute ET_c for weekly or longer time periods, although calculations may proceed on a daily time step. The time-averaged single K_c is used for planning studies and irrigation system design where the averaged effects of soil wetting are acceptable and relevant.

This is the case for surface irrigation and set sprinkler systems where the time interval between successive irrigation is of several days, often ten days or more. For typical irrigation management, the time-averaged single K_c is valid.

Dual Crop Coefficient Approach (Kcb + Ke)

In the dual crop coefficient approach, the effects of crop transpiration and soil evaporation are determined separately. Two coefficients are used: the basal crop coefficient (K_{cb}) to describe plant transpiration, and the soil water evaporation coefficient (K_e) to describe evaporation from the soil surface. The single K_c coefficient is replaced by:

$$K_c = K_{cb} + K_e \quad (57)$$

where

K_{cb} basal crop coefficient,

K_e soil water evaporation coefficient.

The basal crop coefficient, K_{cb}, is defined as the ratio of ET_c to ET_o when the soil surface layer is dry but where the average soil water content of the root zone is adequate to sustain full plant transpiration. The K_{cb} represents the baseline potential K_c in the absence of the additional effects of soil wetting by irrigation or precipitation. The soil evaporation coefficient, K_e, describes the evaporation component from the soil surface. If the soil is wet following rain or irrigation, K_e may be large.

However, the sum of K_{cb} and K_e can never exceed a maximum value, $K_{c\ max}$, determined by the energy available for evapotranspiration at the soil surface. As the soil surface becomes drier, K_e becomes smaller and falls to zero when no water is left for evaporation. The estimation of K_e requires a daily water balance computation for the calculation of the soil water content remaining in the upper topsoil. The dual coefficient approach requires more numerical calculations than the procedure using the single time-averaged K_c coefficient. The dual procedure is best for real time irrigation scheduling, for soil water balance computations, and for research studies where effects of day-to-day variations in soil surface wetness and the resulting impacts on daily ET_c, the soil water profile, and deep percolation fluxes are important. This is the case for high frequency irrigation with microirrigation systems or lateral move systems such as centre pivots and linear move systems.

Crop Coefficient Curve

After the selection of the calculation approach, the determination of the lengths for the crop growth stages and the corresponding crop coefficients, a crop coefficient curve can be constructed. The curve represents the changes in the crop coefficient over the length of the growing season. The shape of the curve represents the changes in the vegetation and ground cover during plant development and maturation that affect the ratio of ET_c to ET_o. From the curve, the K_c factor and hence ET_c can be derived for any period within the growing season.

Single Crop Coefficient

Shortly after the planting of annuals or shortly after the initiation of new leaves for perennials, the value for K_c is small, often less than 0.4. The K_c begins to increase from the initial K_c value, $K_{c\ ini}$, at the beginning of rapid plant development and reaches a maximum value, $K_{c\ mid}$, at the time of maximum or near maximum plant development.

During the late season period, as leaves begin to age and senesce due to natural or cultural practices, the K_c begins to decrease until it reaches a lower value at the end of the growing period equal to $K_{c\ end}$.

Dual Crop Coefficient

The single 'time-averaged' K_c curve incorporates averaged wetting effects into the K_c factor. The value for $K_{c\ mid}$ is relatively constant for most growing and cultural conditions. However, the values for $K_{c\ ini}$ and $K_{c\ end}$ can vary considerably on a daily basis, depending on the frequency of wetting by irrigation and rainfall. The dual crop coefficient approach calculates the actual increases in K_c for each day as a function of plant development and the wetness of the soil surface.

As the single K_c coefficient includes averaged effects of evaporation from the soil, the basal crop coefficient, K_{cb} describing only plant transpiration, lies below the K_c value. The largest difference between K_c and K_{cb} is found in the initial growth stage where evapotranspiration is predominantly in the form of soil evaporation and crop transpiration is still small.

Because crop canopies are near or at full ground cover during the mid-season stage, soil evaporation beneath the canopy has less effect on crop evapotranspiration and the value for K_{cb} in the mid-season stage will be nearly the same as K_c. Depending on the ground cover, the basal crop coefficient during the mid-season may be only 0.05-0.10 lower than the K_c value. Depending on the frequency with which the crop is irrigated during the late season stage, K_{cb} will be similar to (if infrequently irrigated) or less than the K_c value.

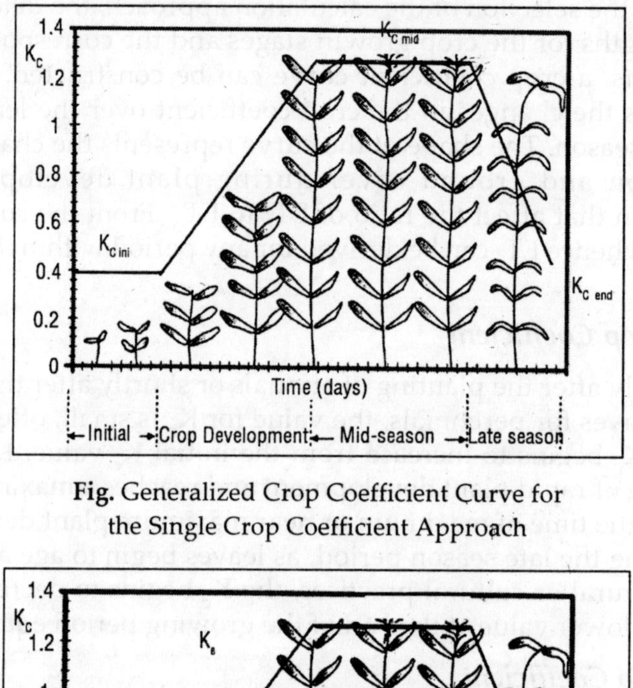

Fig. Generalized Crop Coefficient Curve for the Single Crop Coefficient Approach

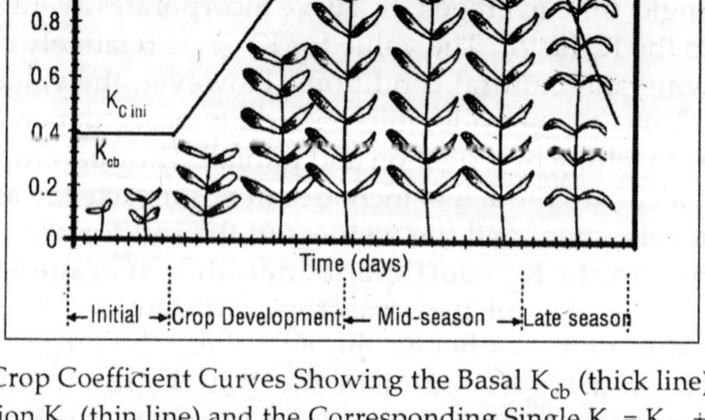

Fig. Crop Coefficient Curves Showing the Basal K_{cb} (thick line), Soil Evaporation K_e (thin line) and the Corresponding Single $K_c = K_{cb} + K_e$ curve (dashed line)

The K_{cb} curve represents the minimum K_c for conditions of adequate soil water and dry soil surface. The K_e 'spikes' in the figure represent increased evaporation when precipitation or irrigation has wetted the soil surface and has temporarily increased total ET_c.

These wet soil evaporation spikes decrease as the soil surface layer dries. The spikes generally reach a maximum value of 1.0-1.2, depending on the climate, the magnitude of the wetting event and the portion of soil surface wetted.

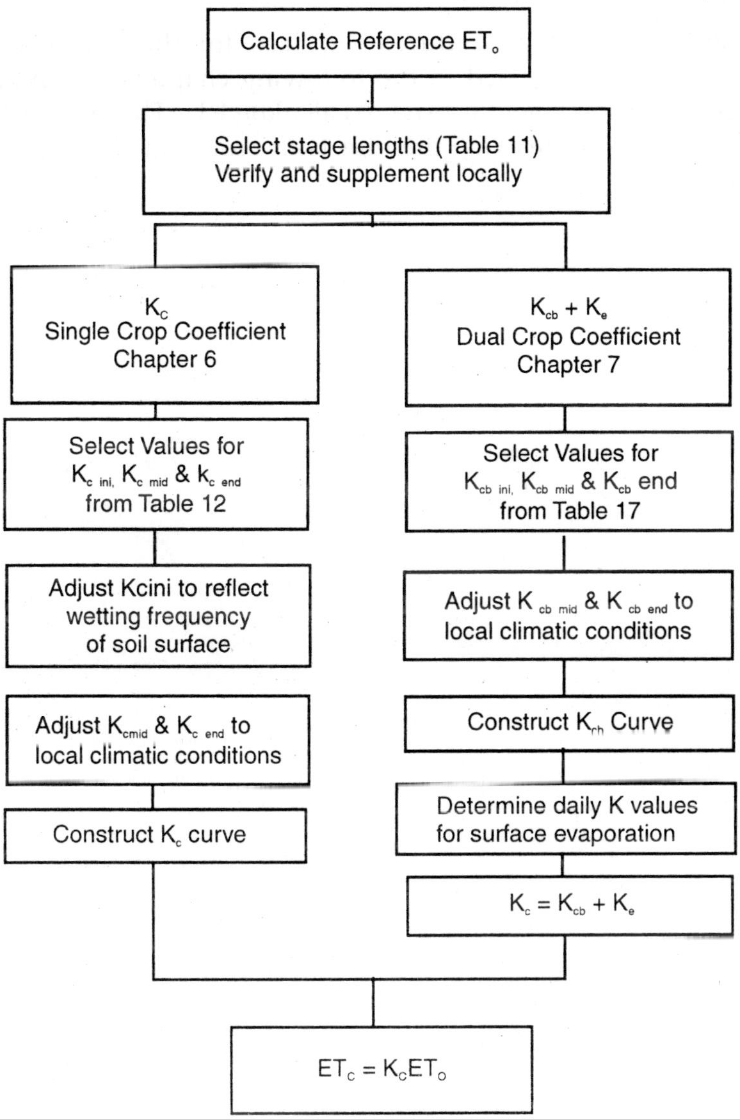

Fig. General Procedure for Calculating ET_c

Summed together, the values for K_{cb} and for K_e represent the single crop coefficient, K_c. The total K_c curve, illustrates the effect of averaging $K_{cb} + K_e$ over time and is displayed as a 'smoothed' curve. It is this smoothed curve that is represented by the single K_c calculation procedure. The K_c curve lies above the K_{cb} curve, with potentially large differences during the initial and development stages, depending on the frequency of soil wetting.

Flow Chart of the Calculations

The calculation procedures required for the crop coefficient approaches are developed in the following chapters. A single time-averaged crop coefficient is used to calculate ET_c. The approach using two coefficients to describe the effects of crop and soil separately is outlined.

Chapter 2

FAO Penman-Monteith Equation

This chapter introduces the user to the need to standardize one method to compute reference evapotranspiration (ET_0) from meteorological data. The FAO Penman-Monteith method is recommended as the sole ET_0 method for determining reference evapotranspiration.

The method, its derivation, the required meteorological data and the corresponding definition of the reference surface are described in this chapter.

NEED FOR A STANDARD ET_0 METHOD

A large number of more or less empirical methods have been developed over the last 50 years by numerous scientists and specialists worldwide to estimate evapotranspiration from different climatic variables. Relationships were often subject to rigorous local calibrations and proved to have limited global validity.

Testing the accuracy of the methods under a new set of conditions is laborious, time-consuming and costly, and yet evapotranspiration data are frequently needed at short notice for project planning or irrigation scheduling design. To meet this need, guidelines were developed and published in the FAO Irrigation and Drainage Paper 'Crop water requirements'

To accommodate users with different data availability, four methods were presented to calculate the reference crop evapotranspiration (ET_0): the Blaney-Criddle, radiation, modified Penman and pan evaporation methods. The modified Penman method was considered to offer the best results with minimum possible error in relation to a living grass reference crop. It was expected that the pan method would give acceptable estimates, depending on the location of the pan.

The radiation method was suggested for areas where available climatic data include measured air temperature and sunshine, cloudiness or radiation, but not measured wind speed and air humidity.

Finally, the publication proposed the use of the Blaney-Criddle method for areas where available climatic data cover air temperature data only.

These climatic methods to calculate ET_o were all calibrated for ten-day or monthly calculations, not for daily or hourly calculations. The Blaney-Criddle method was recommended for periods of one month or longer. For the pan method it was suggested that calculations should be done for periods of ten days or longer. Users have not always respected these conditions and calculations have often been done on daily time steps.

Advances in research and the more accurate assessment of crop water use have revealed weaknesses in the methodologies. Numerous researchers analysed the performance of the four methods for different locations.

Although the results of such analyses could have been influenced by site or measurement conditions or by bias in weather data collection, it became evident that the proposed methods do not behave the same way in different locations around the world.

Deviations from computed to observed values were often found to exceed ranges indicated by FAO. The modified Penman was frequently found to overestimate ET_o, even by up to 20% for low evaporative conditions. The other FAO recommended equations showed variable adherence to the reference crop evapotranspiration standard of grass.

To evaluate the performance of these and other estimation procedures under different climatological conditions, a major study was undertaken under the auspices of the Committee on Irrigation Water Requirements of the American Society of Civil Engineers (ASCE). The ASCE study analysed the performance of 20 different methods, using detailed procedures to assess the validity of the methods compared to a set of carefully screened lysimeter data from 11 locations with variable climatic conditions.

The study proved very revealing and showed the widely varying performance of the methods under different climatic conditions. In a parallel study commissioned by the European Community, a consortium of European research institutes evaluated the performance of various evapotranspiration methods using data from different lysimeter studies in Europe.

The studies confirm the overestimation of the modified Penman introduced in FAO Irrigation and Drainage Paper and the variable performance of the different methods depending on their adaptation to local conditions. The comparative studies may be summarized as follows:

- The Penman methods may require local calibration of the wind function to achieve satisfactory results.
- The radiation methods show good results in humid climates where the aerodynamic term is relatively small, but performance in arid conditions is erratic and tends to underestimate evapotranspiration.
- Temperature methods remain empirical and require local calibration in order to achieve satisfactory results. A possible exception is the 1985 Hargreaves' method which has shown reasonable ET_0 results with a global validity.
- Pan evapotranspiration methods clearly reflect the shortcomings of predicting crop evapotranspiration from open water evaporation. The methods are susceptible to the microclimatic conditions under which the pans are operating and the rigour of station maintenance. Their performance proves erratic.
- The relatively accurate and consistent performance of the Penman-Monteith approach in both arid and humid climates has been indicated in both the ASCE and European studies.

The analysis of the performance of the various calculation methods reveals the need for formulating a standard method for the computation of ET_0.

The FAO Penman-Monteith method is recommended as the sole standard method. It is a method with strong likelihood of correctly predicting ET_0 in a wide range of locations and climates and has provision for application in data-short situations. The use of older FAO or other reference ET methods is no longer encouraged.

FORMULATION OF THE PENMAN-MONTEITH EQUATION

Penman-Monteith Equation

In 1948, Penman combined the energy balance with the mass transfer method and derived an equation to compute the evaporation from an open water surface from standard climatological records of sunshine, temperature, humidity and wind speed. This so-called combination method was further developed by many researchers and extended to cropped surfaces by introducing resistance factors.

The resistance nomenclature distinguishes between aerodynamic resistance and surface resistance factors. The surface resistance parameters are often combined into one parameter, the 'bulk' surface resistance parameter which operates in series with the aerodynamic resistance.

The surface resistance, r_s, describes the resistance of vapour flow through stomata openings, total leaf area and soil surface. The aerodynamic resistance, r_a, describes the resistance from the vegetation upward and involves friction from air flowing over vegetative surfaces.

Although the exchange process in a vegetation layer is too complex to be fully described by the two resistance factors, good correlations can be obtained between measured and calculated evapotranspiration rates, especially for a uniform grass reference surface.

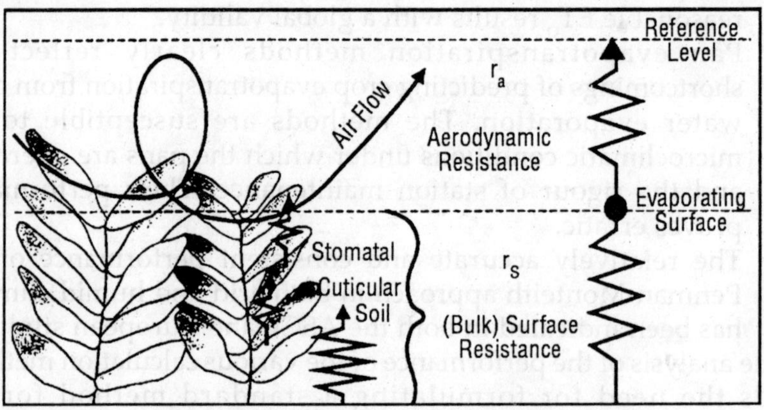

Fig. Simplified Representation of the (Bulk) Surface and Aerodynamic Resistances for Water Vapour Flow

The Penman-Monteith form of the Combination Equation is:

$$\lambda ET = \frac{\Delta(R_n - G) + P_a C_P \frac{(e_s - e_a)}{r_a}}{\Delta + \gamma\left(1 + \frac{r_s}{r_a}\right)}$$

where R_n is the net radiation, G is the soil heat flux, $(e_s - e_a)$ represents the vapour pressure deficit of the air, r $_a$ is the mean air density at constant pressure, c_p is the specific heat of the air, D represents the slope of the saturation vapour pressure temperature relationship, g is the psychrometric constant, and r_s and r_a are the (bulk) surface and aerodynamic resistances.

The Penman-Monteith approach as formulated above includes all parameters that govern energy exchange and corresponding latent heat flux (evapotranspiration) from uniform expanses of vegetation. Most of the parameters are measured or can be readily calculated from weather data. The equation can be utilized for the direct calculation of any crop evapotranspiration as the surface and aerodynamic resistances are crop specific.

Aerodynamic Resistance (r_a)

The transfer of heat and water vapour from the evaporating surface into the air above the canopy is determined by the aerodynamic resistance:

$$r_a = \frac{\text{In}\left[\frac{Z_m - d}{Z_{om}}\right]\text{In}\left[\frac{Z_h - d}{Z_{oh}}\right]}{K^2 u_z}$$

where r_a aerodynamic resistance [s m^{-1}]
z_m height of wind measurements [m]
z_h height of humidity measurements [m]
δ zero plane displacement height [m]
z_{om} roughness length governing momentum transfer [m]
z_{oh} roughness length governing transfer of heat and vapour [m]
k von Karman's constant, 0.41 [–]
u_z wind speed at height z [m s^{-1}].

The equation is restricted for neutral stability conditions, i.e., where temperature, atmospheric pressure, and wind velocity distributions follow nearly adiabatic conditions (no heat exchange). The application of the equation for short time periods (hourly or less) may require the inclusion of corrections for stability. However, when predicting ET_o in the well-watered reference surface, heat exchanged is small, and therefore stability correction is normally not required.

Many studies have explored the nature of the wind regime in plant canopies. Zero displacement heights and roughness lengths have to be considered when the surface is covered by vegetation. The factors depend upon the crop height and architecture. Several empirical equations for the estimate of d, z_{om} and z_{oh} have been developed.

(Bulk) Surface Resistance (r_s)

The 'bulk' surface resistance describes the resistance of vapour flow through the transpiring crop and evaporating soil surface. Where the vegetation does not completely cover the soil, the resistance factor should indeed include the effects of the evaporation from the soil surface.

If the crop is not transpiring at a potential rate, the resistance depends also on the water status of the vegetation. An acceptable approximation to a much more complex relation of the surface resistance of dense full cover vegetation is:

The aerodynamic resistance for a grass reference surface. For a wide range of crops the zero plane displacement height, δ [m], and the roughness length governing momentum transfer, z_{om} [m], can be estimated from the crop height h [m] by the following equations:

$$d = 2/3\ h$$

$$z_{om} = 0.123\ h$$

The roughness length governing transfer of heat and vapour, z_{oh} [m], can be approximated by:

$$z_{oh} = 0.1\ z_{om}$$

Assuming a constant crop height of 0.12 m and a standardized height for wind speed, temperature and humidity at 2 m ($z_m = z_h = 2$ m), the aerodynamic resistance r_a [s m^{-1}] for the grass reference surface becomes:

$$r_a = \frac{\ln\left[\frac{2-2/3(0.12)}{0.123(0.12)}\right]\ln\left[\frac{2-2/3(0.12)}{(0.1)0.123(0.12)}\right]}{(0.41)^2 u_2}$$

where u_2 is the wind speed [m s^{-1}] at 2 m.

$$r_s = \frac{r_1}{LAI_{active}}$$

where r_s (bulk) surface resistance [s m^{-1}],
r_l bulk stomatal resistance of the well-illuminated leaf [s m^{-1}],
LAI_{active} active (sunlit) leaf area index [m^2 (leaf area) m^{-2} (soil surface)].

The Leaf Area Index (LAI), a dimensionless quantity, is the leaf area (upper side only) per unit area of soil below it. It is expressed as m^2 leaf area per m^2 ground area.

The active LAI is the index of the leaf area that actively contributes to the surface heat and vapour transfer. It is generally the upper, sunlit portion of a dense canopy.

The LAI values for various crops differ widely but values of 3-5 are common for many mature crops. For a given crop, green LAI changes throughout the season and normally reaches its maximum before or at flowering. LAI further depends on the plant density and the crop variety.

The bulk stomatal resistance, r_l, is the average resistance of an individual leaf. This resistance is crop specific and differs among crop varieties and crop management. It usually increases as the crop ages and begins to ripen. There is, however, a lack of consolidated information on changes in r_l over time for the different crops.

The information available in the literature on stomatal conductance or resistance is often oriented toward physiological or ecophysiological studies.

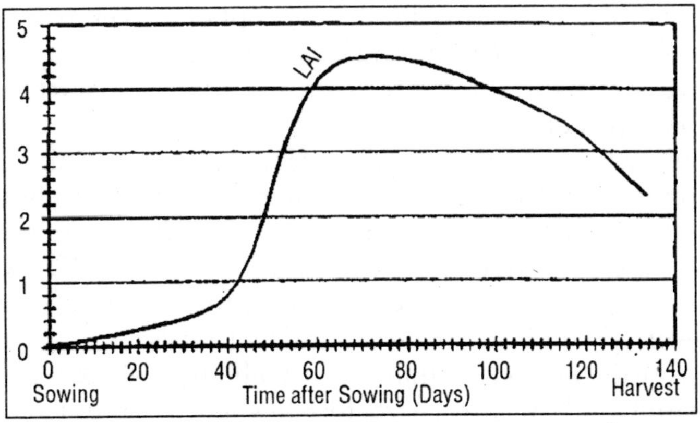

Fig. Typical Presentation of the Variation in the Active (green) Leaf Area Index over the Growing Season for a Maize Crop

The stomatal resistance, r_l, is influenced by climate and by water availability. However, influences vary from one crop to another and different varieties can be affected differently. The resistance increases when the crop is water stressed and the soil water availability limits crop evapotranspiration.

Some studies indicate that stomatal resistance is influenced to some extent by radiation intensity, temperature, and vapour pressure deficit. The derivation of the surface resistance for the grass reference surface is presented.

The (bulk) surface resistance for a grass reference crop

A general equation for LAI_{active} is:

$$LAI_{active} = 0.5\ LAI$$

which takes into consideration the fact that generally only the upper half of dense clipped grass is actively contributing to the surface heat and vapour transfer. For clipped grass a general equation for LAI is:

$$LAI = 24\ h$$

where h is the crop height [m].

The stomatal resistance, r_l, of a single leaf has a value of about 100 s m^{-1} under well-watered conditions. By assuming a crop height of 0.12 m, the surface resistance, r_s [s m^{-1}], for the grass reference surface becomes

$$r_s = \frac{100}{0.5(24)(0.12)} \approx 70 sm^{-1}$$

Reference Surface

To obviate the need to define unique evaporation parameters for each crop and stage of growth, the concept of a reference surface was introduced. Evapotranspiration rates of the various crops are related to the evapotranspiration rate from the reference surface (ET_0) by means of crop coefficients.

In the past, an open water surface has been proposed as a reference surface. However, the differences in aerodynamic, vegetation control and radiation characteristics present a strong challenge in relating ET to measurements of free water evaporation. Relating ET_0 to a specific crop has the advantage of incorporating the biological and physical processes involved in ET from cropped surfaces.

Grass, together with alfalfa, is a well-studied crop regarding its aerodynamic and surface characteristics and is accepted worldwide as a reference surface. Because the resistance to diffusion of vapour strongly depends on crop height, ground cover, LAI and soil moisture conditions, the characteristics of the reference crop should be well defined and fixed.

Changes in crop height result in variations in the roughness and LAI. Consequently, the associated canopy and aerodynamic resistances will vary appreciably with time. Moreover, water stress and the degree of ground cover have an effect on the resistances and also on the albedo.

To avoid problems of local calibration which would require demanding and expensive studies, a hypothetical grass reference has been selected.

Difficulties with a living grass reference result from the fact that the grass variety and morphology can significantly affect the evapotranspiration rate, especially during peak water use.

- Large differences may exist between warm-season and cool-season grass types. Cool-season grasses have a lower degree of stomatal control and hence higher rates of evapotranspiration. It may be difficult to grow cool-season grasses in some arid, tropical climates.

The FAO Expert Consultation on Revision of FAO Methodologies for Crop Water Requirements accepted the following unambiguous definition for the reference surface:

A hypothetical reference crop with an assumed crop height of 0.12 m, a fixed surface resistance of 70 s m^{-1} and an albedo of 0.23."

The reference surface closely resembles an extensive surface of green grass of uniform height, actively growing, completely shading the ground and with adequate water. The requirements that the grass surface should be extensive and uniform result from the assumption that all fluxes are one-dimensional upwards.

The FAO Penman-Monteith method is selected as the method by which the evapotranspiration of this reference surface (ET_0) can be unambiguously determined, and as the method which provides consistent ET_o values in all regions and climates.

FAO PENMAN-MONTEITH EQUATION

Equation

A consultation of experts and researchers was organized by FAO in May 1990, in collaboration with the International Commission for Irrigation and Drainage and with the World Meteorological Organization, to review the FAO methodologies on crop water requirements and to advise on the revision and update of procedures.

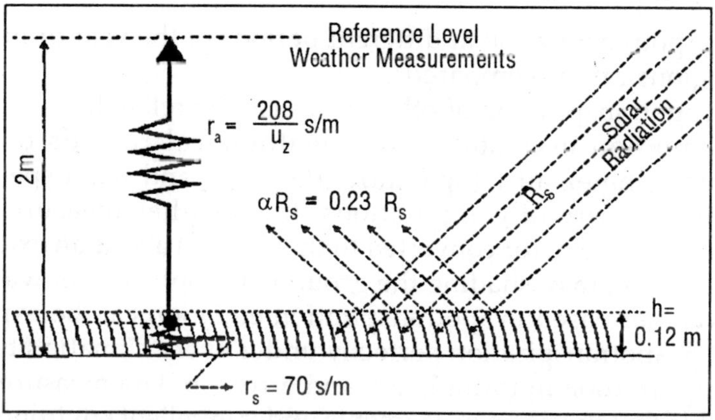

Fig. Characteristics of the Hypothetical Reference Crop

The panel of experts recommended the adoption of the Penman-Monteith combination method as a new standard for reference evapotranspiration and advised on procedures for calculation of the various parameters.

By defining the reference crop as a hypothetical crop with an assumed height of 0.12 m having a surface resistance of 70 s m^{-1} and an albedo of 0.23, closely resembling the evaporation of an extension surface of green grass of uniform height, actively growing and adequately watered, the FAO Penman-Monteith method was developed. The method overcomes shortcomings of the previous FAO Penman method and provides values more consistent with actual crop water use data worldwide. From the original Penman-Monteith equation and the equations of the aerodynamic and surface resistance, the FAO Penman-Monteith method to estimate ET_o can be derived:

$$ET_o = \frac{0.408\Delta(R_n - G) + \gamma\frac{900}{T+273}u_2(e_s - e_a)}{\Delta + \gamma(1 + 0.34u_2)}$$

where ET_o reference evapotranspiration [mm day^{-1}]
R_n net radiation at the crop surface [MJ m^{-2} day^{-1}]
G soil heat flux density [MJ m^{-2} day^{-1}]
T mean daily air temperature at 2 m height [°C]
u_2 wind speed at 2 m height [m s^{-1}]
e_s saturation vapour pressure [kPa]
e_a actual vapour pressure [kPa]
$e_s - e_a$ saturation vapour pressure deficit [kPa]
D slope vapour pressure curve [kPa °C^{-1}]
g psychrometric constant [kPa °C^{-1}].

The reference evapotranspiration, ET_o, provides a standard to which:

- Evapotranspiration at different periods of the year or in other regions can be compared.
- Evapotranspiration of other crops can be related.

The equation uses standard climatological records of solar radiation (sunshine), air temperature, humidity and wind speed. To ensure the integrity of computations, the weather measurements should be made at 2 m (or converted to that height) above an extensive surface of green grass, shading the ground and not short of water.

No weather-based evapotranspiration equation can be expected to predict evapotranspiration perfectly under every climatic situation due to simplification in formulation and errors in data measurement. It is probable that precision instruments under excellent environmental and biological management conditions will show the FAO Penman-Monteith equation to deviate at times from true measurements of grass ET_o. However, the Expert Consultation agreed to use the hypothetical reference definition of the FAO Penman-Monteith equation as the definition for grass ET_o when deriving and expressing crop coefficients.

It is important, when comparing the FAO Penman-Monteith equation to ET_o measurements, that the full Penman-Monteith equation and associated equations for r_a and r_s be used to enable accounting for variation in ET due to variation in height of the grass measured.

Variations in measurement height can significantly change LAI, d and z_{om} and the corresponding ET_o measurement and predicted value. When evaluating results, it should be noted that local environmental and management factors, such as watering frequency, also affect ET_o observations.

The FAO Penman-Monteith equation is a close, simple representation of the physical and physiological factors governing the evapotranspiration process. By using the FAO Penman-Monteith definition for ET_o, one may calculate crop coefficients at research sites by relating the measured crop evapotranspiration (ET_c) with the calculated ET_o, i.e.,

$$K_c = ET_c/ET_o.$$

In the crop coefficient approach, differences in the crop canopy and aerodynamic resistance relative to the hypothetical reference crop are accounted for within the crop coefficient. The K_c factor serves as an aggregation of the physical and physiological differences between crops and the reference definition.

DATA

Apart from the site location, the FAO Penman-Monteith equation requires air temperature, humidity, radiation and wind speed data for daily, weekly, ten-day or monthly calculations. The computation of all data required for the calculation of the reference evapotranspiration. It is important to verify the units in which the weather data are reported.

Location

Altitude above sea level (m) and latitude (degrees north or south) of the location should be specified. These data are needed to adjust some weather parameters for the local average value of atmospheric pressure (a function of the site elevation above mean sea level) and to compute extraterrestrial radiation (R_a) and, in some cases, daylight hours (N). In the calculation procedures for R_a and N, the latitude is expressed in radian (i.e., decimal degrees times π /180).

Derivation of the FAO Penman-Monteith equation for the hypothetical grass reference crop. With standardized height for wind speed, temperature and humidity measurements at 2 m ($z_m = z_h = 2$ m) and the crop height h = 0.12 m, the aerodynamic and surface resistances become:

$r_a - 208/u_2$ s m^{-1}, (with u_2 wind speed at 2 m height)

$r_s = 70$ s m^{-1}

$$(1 + r_s/r_a) - (1 + 0.34\ u_2)$$

R_n and G is energy available per unit area and expressed in MJ m^{-2} day^{-1}. To convert the energy units for radiation to equivalent water depths (mm) the latent heat of vaporization, l is used as a conversion factor. The conversion from energy values to equivalent depths of water or vice versa is given by:

$$\text{Raditior}[\text{mm day}^{-1}] \approx \frac{\text{Raditior}(\text{MJM}^{-2}\text{day}^{-1}]}{2.45} = 0.408\,\text{Raditior}[\text{MJM}^{-2}\ \text{day}^{-1}$$

By substituting c_p with a rearrangement of Equation:

$$C_P = \frac{\gamma\varepsilon\lambda}{P}$$

and considering the ideal gas law for r a:

$$P_a = \frac{P}{T_{KV}R}$$

where T_{Kv} the virtual temperature, may be substituted by:

$$T_{Kv} = 1.01(T + 273)$$

results in:

$$\frac{C_P P_a}{r_a} = \frac{\gamma\varepsilon\lambda}{1.01(T+273)R(208)} U2[MJ\,m^{-2}\,°C^{-1}\,day^{-1}]$$

Where, c_p specific heat at constant pressure [MJ $kg^{-1}°C^{-1}$],
r_a mean air density at constant pressure [kg m^{-3}],
r_a aerodynamic resistance [s m^{-1}],
g psychrometric constant [kPa$°C^{-1}$],
e ratio molecular weight of water vapour/dry air = 0.622,
l latent heat of vaporization [MJ kg^{-1}],
u_2 wind speed at 2 m [m s^{-1}],
R specific gas constant = 0.287 kJ kg^{-1} K^{-1},
T air temperature [°C],
P atmospheric pressure [kPa],

$$86400 = \frac{\gamma(0.622)\lambda}{1.01(T+273)R(208)} u_2[MJ\,m^{-2}\,°C^{-1}\,day^{-1}]$$

or, when divided by $\lambda(\lambda = 2.45)$

$$\approx \gamma\frac{900}{T+273} u_2[mm\,°C^{-1}\,day^{-1}]$$

A positive value is used for the northern hemisphere and a negative value for the southern hemisphere.

Temperature

The (average) daily maximum and minimum air temperatures in degrees Celsius (°C) are required. Where only (average) mean daily temperatures are available, the calculations can still be executed but some underestimation of ET_o will probably occur due to the non-linearity of the saturation vapour pressure - temperature relationship.

Using mean air temperature instead of maximum and minimum air temperatures yields a lower saturation vapour pressure e_s, and hence a lower vapour pressure difference $(e_s - e_a)$, and a lower reference evapotranspiration estimate.

Humidity

The (average) daily actual vapour pressure, e_a, in kilopascals (kPa) is required. The actual vapour pressure, where not available, can be derived from maximum and minimum relative humidity (%), psychrometric data (dry and wet bulb temperatures in °C) or dewpoint temperature (°C).

Radiation

The (average) daily net radiation expressed in megajoules per square metre per day (MJ m^{-2} day^{-1}) is required. These data are not commonly available but can be derived from the (average) shortwave radiation measured with a pyranometer or from the (average) daily actual duration of bright sunshine (hours per day) measured with a (Campbell-Stokes) sunshine recorder.

Wind Speed

The (average) daily wind speed in metres per second (m s^{-1}) measured at 2 m above the ground level is required. It is important to verify the height at which wind speed is measured, as wind speeds measured at different heights above the soil surface differ.

MISSING CLIMATIC DATA

Situations might occur where data for some weather variables are missing. The use of an alternative ET_o calculation procedure, requiring only limited meteorological parameters, should generally be avoided. It is recommended that one calculate ET_o using the standard FAO Penman-Monteith method after resolving the specific problem of the missing data.

Procedures for estimating missing climatic data. Differences between ET_o values obtained with the FAO Penman-Monteith equation with, on the one hand, a limited data set and, on the other hand, a full data set, are expected to be smaller than or of similar magnitude to the differences resulting from the use of an alternative ET_o equation.

Even where the data set contains only maximum and minimum air temperature it is still possible to obtain reasonable estimates of ten-day or monthly ET_o with the FAO Penman-Monteith equation. Radiation data can be derived from the air temperature difference, or, along with wind speed and humidity data, can be imported from a nearby weather station.

Humidity data can also be estimated from daily minimum air temperature. After evaluating the validity of the use of data from another station, ten-day or monthly estimates of ET_o can be calculated.

The procedures for estimating missing data should be validated at the regional level. This can be done for weather stations with full data sets by comparing ET_o calculated with full and with limited data sets. The ratio should be close to one. Where the ratio deviates significantly from one, the ratio can be used as a correction factor for estimates made with the limited data set.

Where the standard error of estimate exceeds 20% of the mean ET_o, a sensitivity analysis should be performed to determine causes (and limits) for the method utilized to import the missing data. A validation should be completed for each month and variable, for the monthly as well as for the daily estimates.

Chapter 3

Meteorological Data

The methods for calculating evapotranspiration from meteorological data require various climatological and physical parameters. Some of the data are measured directly in weather stations. Other parameters are related to commonly measured data and can be derived with the help of a direct or empirical relationship.

This chapter discusses the source, measurement and computation of all data required for the calculation of the reference evapotranspiration by means of the FAO Penman-Monteith method. Different examples illustrate the various calculation procedures. Appropriate procedures for estimating missing data are also provided.

METEOROLOGICAL FACTORS DETERMINING ET

The meteorological factors determining evapotranspiration are weather parameters which provide energy for vaporization and remove water vapour from the evaporating surface. The principal weather parameters to consider are presented below.

Solar Radiation

The evapotranspiration process is determined by the amount of energy available to vaporize water. Solar radiation is the largest energy source and is able to change large quantities of liquid water into water vapour. The potential amount of radiation that can reach the evaporating surface is determined by its location and time of the year. Due to differences in the position of the sun, the potential radiation differs at various latitudes and in different seasons.

The actual solar radiation reaching the evaporating surface depends on the turbidity of the atmosphere and the presence of clouds which reflect and absorb major parts of the radiation. When assessing the effect of solar radiation on evapotranspiration, one should also bear in mind that not all available energy is used to vaporize water. Part of the solar energy is used to heat up the atmosphere and the soil profile.

AIR TEMPERATURE

The solar radiation absorbed by the atmosphere and the heat emitted by the earth increase the air temperature. The sensible heat of the surrounding air transfers energy to the crop and exerts as such a controlling influence on the rate of evapotranspiration. In sunny, warm weather the loss of water by evapotranspiration is greater than in cloudy and cool weather.

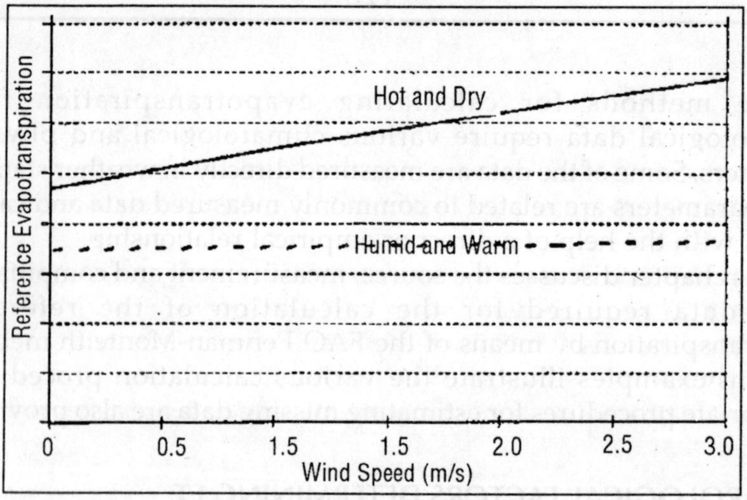

Fig. Illustration of the Effect of Wind Speed on Evapotranspiration in Hot-dry and Humid-warm Weather Conditions

Air Humidity

While the energy supply from the sun and surrounding air is the main driving force for the vaporization of water, the difference between the water vapour pressure at the evapotranspiring surface and the surrounding air is the determining factor for the vapour removal. Well-watered fields in hot dry arid regions consume large amounts of water due to the abundance of energy and the desiccating power of the atmosphere.

In humid tropical regions, notwithstanding the high energy input, the high humidity of the air will reduce the evapotranspiration demand. In such an environment, the air is already close to saturation, so that less additional water can be stored and hence the evapotranspiration rate is lower than in arid regions.

Wind Speed

The process of vapour removal depends to a large extent on wind and air turbulence which transfers large quantities of air over the evaporating surface.

When vaporizing water, the air above the evaporating surface becomes gradually saturated with water vapour. If this air is not continuously replaced with drier air, the driving force for water vapour removal and the evapotranspiration rate decreases.

The combined effect of climatic factors affecting evapotranspiration for two different climatic conditions. The evapotranspiration demand is high in hot dry weather due to the dryness of the air and the amount of energy available as direct solar radiation and latent heat.

Under these circumstances, much water vapour can be stored in the air while wind may promote the transport of water allowing more water vapour to be taken up.

On the other hand, under humid weather conditions, the high humidity of the air and the presence of clouds cause the evapotranspiration rate to be lower. The effect on evapotranspiration of increasing wind speeds for the two different climatic conditions is illustrated by the slope of the curves.

The drier the atmosphere, the larger the effect on ET and the greater the slope of the curve. For humid conditions, the wind can only replace saturated air with slightly less saturated air and remove heat energy. Consequently, the wind speed affects the evapotranspiration rate to a far lesser extent than under arid conditions where small variations in wind speed may result in larger variations in the evapotranspiration rate.

ATMOSPHERIC PARAMETERS

Several relationships are available to express climatic parameters. The effect of the principal weather parameters on evapotranspiration can be assessed with the help of these equations. Some of the relationships require parameters which express a specific characteristic of the atmosphere. Before studying the four principal weather parameters, some atmospheric parameters will be discussed.

Atmospheric Pressure (P)

The atmospheric pressure, P, is the pressure exerted by the weight of the earth's atmosphere. Evaporation at high altitudes is promoted due to low atmospheric pressure as expressed in the psychrometric constant. The effect is, however, small and in the calculation procedures, the average value for a location is sufficient. A simplification of the ideal gas law, assuming 20°C for a standard atmosphere, can be employed to calculate P:

$$P = 101.3\left(\frac{293 - 0.0065z}{293}\right)^{526} \quad (7)$$

where, P atmospheric pressure [kPa]
z elevation above sea level [m].

Latent Heat of Vaporization (λ)

The latent heat of vaporization, l, expresses the energy required to change a unit mass of water from liquid to water vapour in a constant pressure and constant temperature process. The value of the latent heat varies as a function of temperature.

At a high temperature, less energy will be required than at lower temperatures. As l varies only slightly over normal temperature ranges a single value of 2.45 MJ kg^{-1} is taken in the simplification of the FAO Penman-Monteith equation. This is the latent heat for an air temperature of about 20°C.

Psychrometric Constant (γ)

The psychrometric constant, γ, is given by:

$$\gamma = \frac{C_p P}{\varepsilon\lambda} = 0.665 \times 10^{-3} P$$

where, g psychrometric constant [kPa $°C^{-1}$]
P atmospheric pressure [kPa]
l latent heat of vaporization, 2.45 [MJ kg^{-1}]
c_p specific heat at constant pressure, 1.013 10^{-3} [MJ kg^{-1} $°C^{-1}$]
e ratio molecular weight of water vapour/dry air = 0.622.

The specific heat at constant pressure is the amount of energy required to increase the temperature of a unit mass of air by one degree at constant pressure. Its value depends on the composition of the air, i.e., on its humidity. For average atmospheric conditions a value c_p = 1.013 × 10^{-3} MJ kg^{-1} $°C^{-1}$ can be used. As an average atmospheric pressure is used for each location , the psychrometric constant is kept constant for each location.

Example. Determination of Atmospheric Parameters.

Determine the atmospheric pressure and the psychrometric constant at an elevation of 1800 m.			
With:	z =	1800	m
From Eq. 7:	P = 101.3 [(293 - (0.0065) 1800)/293]$^{5.26}$ =	81.8	kPa
From Eq. 8:	γ = 0.665 10^{-3} (81.8) =	0.054	kPa $°C^{-1}$
The average value of the atmospheric pressure is 81.8 kPa. The psychrometric constant, γ, is 0.054 kPa/°C.			

Air Temperature

Agrometeorology is concerned with the air temperature near the level of the crop canopy. In traditional and modem automatic weather stations the air temperature is measured inside shelters (Stevenson screens or ventilated radiation shields) placed in line with World Meteorological Organization (WMO) standards at 2 m above the ground.

The shelters are designed to protect the instruments from direct exposure to solar heating. The louvered construction allows free air movement around the instruments.

Air temperature is measured with thermometers, thermistors or thermocouples mounted in the shelter. Minimum and maximum thermometers record the minimum and maximum air temperature over a 24-hour period.

Thermographs plot the instantaneous temperature over a day or week. Electronic weather stations often sample air temperature each minute and report hourly averages in addition to 24-hour maximum and minimum values.

Due to the non-linearity of humidity data required in the FAO Penman-Monteith equation, the vapour pressure for a certain period should be computed as the mean between the vapour pressure at the daily maximum and minimum air temperatures of that period.

The daily maximum air temperature (T_{max}) and daily minimum air temperature (T_{min}) are, respectively, the maximum and minimum air temperature observed during the 24-hour period, beginning at midnight. T_{max} and T_{min} for longer periods such as weeks, 10-day's or months are obtained by dividing the sum of the respective daily values by the number of days in the period.

The mean daily air temperature (T_{mean}) is only employed in the FAO Penman-Monteith equation to calculate the slope of the saturation vapour pressure curves (D) and the impact of mean air density (P_a) as the effect of temperature variations on the value of the climatic parameter is small in these cases. For standardization, T_{mean} for 24-hour periods is defined as the mean of the daily maximum (T_{max}) and minimum temperatures (T_{min}) rather than as the average of hourly temperature measurements.

$$T_{mean} = \frac{T_{max} - T_{min}}{2}$$

The temperature is given in degrees Celsius (°C) or Fahrenheit (°F). In some calculation procedures, temperature is required in Kelvin (K), which can be obtained by adding 273.16 to the temperature expressed in degrees Celsius (in practice K = °C + 273.16). The Kelvin and Celsius scale have the same scale interval.

AIR HUMIDITY

Concepts

The water content of the air can be expressed in several ways. In agrometeorology, vapour pressure, dewpoint temperature and relative humidity are common expressions to indicate air humidity.

Vapour Pressure

Water vapour is a gas and its pressure contributes to the total atmospheric pressure. The amount of water in the air is related directly to the partial pressure exerted by the water vapour in the air and is therefore a direct measure of the air water content.

In standard S. I. units, pressure is no longer expressed in centimetre of water, millimetre of mercury, bars, atmosphere, etc., but in pascals (Pa). Conversion factors between various units and Pa.

As a pascal refers to a relatively small force (1 newton) applied on a relatively large surface (1 m^2), multiples of the basic unit are often used. In this handbook, vapour pressure is expressed in kilopascals (kPa = 1000 Pa). When air is enclosed above an evaporating water surface, an equilibrium is reached between the water molecules escaping and returning to the water reservoir.

At that moment, the air is said to be saturated since it cannot store any extra water molecules. The corresponding pressure is called the saturation vapour pressure (e°(T)). The number of water molecules that can be stored in the air depends on the temperature (T). The higher the air temperature, the higher the storage capacity, the higher its saturation vapour pressure. The slope of the curve changes exponentially with temperature. At low temperatures, the slope is small and varies only slightly as the temperature rises. At high temperatures, the slope is large and small changes in T result in large changes in slope. The slope of the saturation vapour pressure curve, D, is an important parameter in describing vaporization and is required in the equations for calculating ET_o from climatic data.

The actual vapour pressure (e_a) is the vapour pressure exerted by the water in the air. When the air is not saturated, the actual vapour pressure will be lower than the saturation vapour pressure. The difference between the saturation and actual vapour pressure is called the vapour pressure deficit or saturation deficit and is an accurate indicator of the actual evaporative capacity of the air.

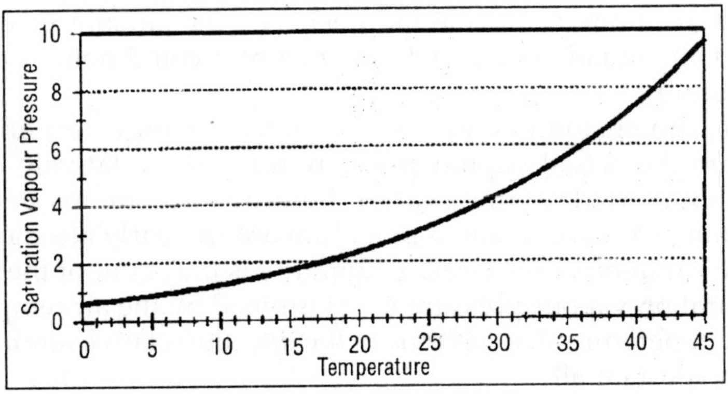

Fig. Saturation Vapour Pressure Shown as a Function of Temperature: e°(T) Curve

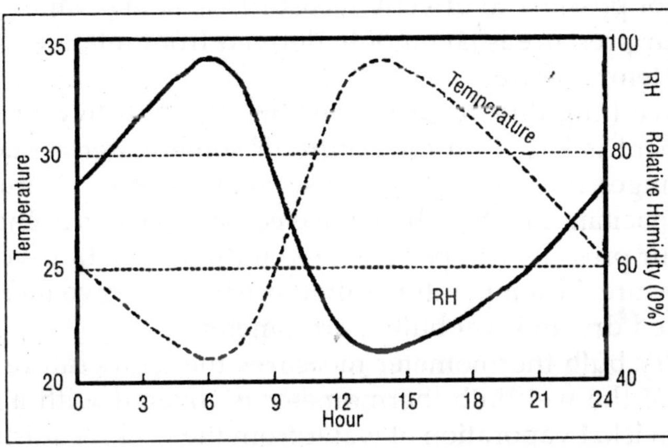

Fig. Variation of the Relative Humidity over 24 hours for a Constant Actual Vapour Pressure of 2.4 kPa

Dewpoint Temperature

The dewpoint temperature is the temperature to which the air needs to be cooled to make the air saturated. The actual vapour pressure of the air is the saturation vapour pressure at the dewpoint temperature, The drier the air, the larger the difference between the air temperature and dewpoint temperature.

Relative Humidity

The relative humidity (RH) expresses the degree of saturation of the air as a ratio of the actual (e_a) to the saturation ($e^\circ(T)$) vapour pressure at the same temperature (T):

$$RH = 100\frac{e_a}{e^o(T)}$$

Relative humidity is the ratio between the amount of water the ambient air actually holds and the amount it could hold at the same temperature.

It is dimensionless and is commonly given as a percentage. Although the actual vapour pressure might be relatively constant throughout the day, the relative humidity fluctuates between a maximum near sunrise and a minimum around early afternoon.

The variation of the relative humidity is the result of the fact that the saturation vapour pressure is determined by the air temperature. As the temperature changes during the day, the relative humidity also changes substantially.

MEASUREMENT

It is not possible to directly measure the actual vapour pressure. The vapour pressure is commonly derived from relative humidity or dewpoint temperature.

Relative humidity is measured directly with hygrometers. The measurement is based on the nature of some material such as hair, which changes its length in response to changes in air humidity, or using a capacitance plate, where the electric capacitance changes with RH. Vapour pressure can be measured indirectly with psychrometers which measure the temperature difference between two thermometers, the so-called dry and wet bulb thermometers.

The dry bulb thermometer measures the temperature of the air. The bulb of the wet bulb thermometer is covered with a constantly saturated wick. Evaporation of water from the wick, requiring energy, lowers the temperature of the thermometer. The drier the air, the larger the evaporative cooling and the larger the temperature drop. The difference between the dry and wet bulb temperatures is called the wet bulb depression and is a measure of the air humidity.

The dewpoint temperature is measured with dewpoint meters. The underlying principle of some types of apparatus is the cooling of the ambient air until dew formation occurs. The corresponding temperature is the dewpoint temperature.

Relative humidity and dewpoint temperature data are notoriously plagued by measurement errors. Measurement error is common for both older hygrothermograph types of instruments and for the more modem electronic instruments. Great care should be made to assess the accuracy and integrity of RH and dewpoint data.

The user is encouraged to always compare computed dewpoint temperatures to daily minimum air temperatures. Frequently, it is better to utilize a dewpoint temperature that is predicted from daily minimum air temperature, rather than to use unreliable relative humidity measurements. The user is encouraged to utilize good judgement in this area.

CALCULATION PROCEDURES

Mean Saturation Vapour Pressure (es)

As saturation vapour pressure is related to air temperature, it can be calculated from the air temperature. The relationship is expressed by:

$$e^{o}(T) = 0.6108 \exp\left[\frac{17.27T}{T + 237.3}\right]$$

where, e°(T) saturation vapour pressure at the air temperature T [kPa]

T air temperature [°C].

exp[..] 2.7183 (base of natural logarithm) raised to the power [..].

Due to the non-linearity of the above equation, the mean saturation vapour pressure for a day, week, decade or month should be computed as the mean between the saturation vapour pressure at the mean daily maximum and minimum air temperatures for that period:

$$e_s = \frac{e^{o}(T_{max}) + e^{o}(T_{min})}{2}$$

Using mean air temperature instead of daily minimum and maximum temperatures results in lower estimates for the mean saturation vapour pressure. The corresponding vapour pressure deficit (a parameter expressing the evaporating power of the atmosphere) will also be smaller and the result will be some underestimation of the reference crop evapotranspiration. Therefore, the mean saturation vapour pressure should be calculated as the mean between the saturation vapour pressure at both the daily maximum and minimum air temperature.

Example. Determination of mean saturation vapour pressure

The daily maximum and minimum air temperature are respectively 24.5 and 15°C. Determine the saturation vapour pressure for that day.			
From Eq. 11	$e^{o}(T_{max}) = 0.6108 \exp[17.27(24.5)/(24.5 + 237.3)]$	3.075	kPa
From Eq. 11	$e^{o}(T_{min}) = 0.6108 \exp[17.27(15)/(15 + 237.3)]$	1.705	kPa
From Eq. 12	$e_s = (3.075 + 1.705)/2$	2.39	kPa
	Note that for temperature 19.75°C (which is T_{mean}), e°(T) =	2.30	kPa
The mean saturation vapour pressure is 2.39 kPa.			

Slope of Saturation Vapour Pressure Curve (Δ)

For the calculation of evapotranspiration, the slope of the

relationship between saturation vapour pressure and temperature, D, is required. The slope of the curve at a given temperature is given by.

$$\Delta = \frac{4098\left[0.6108\exp\left(\frac{17.27T}{T+237.3}\right)\right]}{(T+237.3)^2}$$

where, Δ slope of saturation vapour pressure curve at air temperature T [kPa °C^{-1}]

T is air temperature [°C]

exp[..] 2.7183 (base of natural logarithm) raised to the power [..].

In the FAO Penman-Monteith equation, where Δ occurs in the numerator and denominator, the slope of the vapour pressure curve is calculated using mean air temperature.

Actual Vapour Pressure (e_a) Derived from Dewpoint Temperature

As the dewpoint temperature is the temperature to which the air needs to be cooled to make the air saturated, the actual vapour pressure (e_a) is the saturation vapour pressure at the dewpoint temperature (T_{dew}) [°C], or:

$$e_a = e^o(T_{dew}) = 0.6108\exp\left[\frac{17.27T_{dew}}{T_{dew}+237.3}\right]$$

Actual Vapour Pressure (e_a) Derived from Psychrometric Data

The actual vapour pressure can be determined from the difference between the dry and wet bulb temperatures, the so-called wet bulb depression. The relationship is expressed by the following equation:

$$e_a = e^o(T_{wet}) - \gamma_{psy}(T_{dry} - T_{wet})$$

where, e_a actual vapour pressure [kPa]

$e^o(T_{wet})$ saturation vapour pressure at wet bulb temperature [kPa]

γ_{psy} psychrometric constant of the instrument [kPa °C^{-1}]

T_{dry}-T_{wet} wet bulb depression, with T_{dry} the dry bulb and T_{wet} the wet bulb temperature [°C].

The psychrometric constant of the instrument is given by:

$$\gamma_{psy} = a_{psy}\ P$$

where a_{psy} is a coefficient depending on the type of ventilation of the wet bulb [°C^{-1}], and P is the atmospheric pressure [kPa]. The coefficient a_{psy} depends mainly on the design of the psychrometer and rate of ventilation around the wet bulb. The following values are used

a_{psy} =	0.000662	for ventilated (Asmann type) psychrometers, with an air movement of some 5 m/s,
	0.000800	for natural ventilated psychrometers (about 1 m/s),
	0.001200	for non-ventilated psychrometers installed indoors.

Example. Determination of actual vapour pressure from psychrometric readings

Determine the vapour pressure from the readings of an aspirated psychrometer in a location at an elevation of 1200 m. The temperatures measured by the dry and wet bulb thermometers are 25.6 and 19.5°C respectively.			
From Eq. 7 (Table 2.1), at:	z=	1200	m
Then:	P=	87.9	kPa
From Eq. 11 (Table 2.3), for	T_{wet} =	19.5	°C
Then:	$e^o(T_{wet})$ =	2.267	kPa
Ventilated psychrometer	a_{psy} =	0.000662	$°C^{-1}$
From Eq. 15:	e_a = 2.267 - 0.000662 (87.9) (25.6 - 19.5) =	1.91	kPa
The actual vapour pressure is 1.91 kPa.			

Actual Vapour Pressure (e_a) Derived from Relative Humidity Data

The actual vapour pressure can also be calculated from the relative humidity. Depending on the availability of the humidity data, different equations should be used.

- For RH_{max} and RH_{min}:

$$e_a = \frac{e^o(T_{min})\frac{RH_{max}}{100} + e^o(T_{max})\frac{RH_{min}}{100}}{2}$$

Where, e_a actual vapour pressure [kPa],

$e^o(T_{min})$ saturation vapour pressure at daily minimum temperature [kPa],

$e^o(T_{max})$ saturation vapour pressure at daily maximum temperature [kPa],

RH_{max} maximum relative humidity [%],

RH_{min} minimum relative humidity [%].

For periods of a week, ten days or a month, RH_{max} and RH_{min} are obtained by dividing the sum of the daily values by the number of days in that period.

- For RH_{max}:

When using equipment where errors in estimating RH_{min} can be large, or when RH data integrity are in doubt, then one should use only RH_{max}:

$$e_a = e_o(T_{min})\frac{RH_{max}}{100}$$

- For RH_{mean}:

In the absence of RH_{max} and RH_{min}, another equation can be used to estimate e_a:

$$e_a = \frac{RH_{mean}}{100}\left[\frac{e^o(T_{max}) + e^o(T_{min})}{2}\right]$$

Where RH_{mean} is the mean relative humidity, defined as the average between RH_{max} and RH_{min}. However, Equation is less desirable than are Equations.

Example. Determination of actual vapour pressure from relative humidity

Given the following daily minimum and maximum air temperature and the corresponding relative humidity data:

T_{min} = 18°C and RH_{max} = 82%
T_{max} = 25°C and RH_{min} = 54%

Determine the actual vapour pressure.

From Eq. 11 (Table 2.3), at:	T_{min} =	18	°C
Then:	$e^o(T_{min})$ =	2.064	kPa
From Eq. 11 (Table 2.3), at:	T_{max} =	25	°C
Then:	$e^o(T_{max})$ =	3.168	kPa
From Eq. 17:	e_a = [2.064 (82/100) + 3.168 (54/100)] =	1.70	kPa
Note that when using Eq. 19:	e_a =	1.78	kPa

Vapour Pressure Deficit ($e_s - e_a$)

The vapour pressure deficit is the difference between the saturation (e_s) and actual vapour pressure (e_a) for a given time period. For time periods such as a week, ten days or a month e_s is computed from Equation 12 using the T_{max} and T_{min} averaged over the time period and similarly the e_a is computed with one of the equations 4 to 19, using average measurements over the period. As stated above, using mean air temperature and not T_{max} and T_{min} in Equation 12 results in a lower estimate of e_s, thus in a lower vapour pressure deficit and hence an underestimation of the ET_o. When desired, e_s and e_a for long time periods cal also be calculated as averages of values computed for each day of the period.

Example. Determination of vapour pressure deficit

Determine the vapour pressure deficit with the data of the previous example (Example 5).			
From Example 5:	$e^\circ(T_{min})$ =	2.064	kPa
	$e^\circ(T_{max})$ =	3.168	kPa
	e_a =	1.70	kPa
	$e_s - e_a = (2.064 + 3.168)/2 - 1.70$ =	0.91	kPa
The vapour pressure deficit is 0.91 kPa.			

Calculation sheet for vapour pressure deficit ($e_s - e_a$)

BOX 7. Calculation sheet for vapour pressure deficit ($e_s - e_a$)				
Saturation vapour pressure: e_s (Eq. 11 or Table 2.3)				
T_{max}		°C	$e^\circ(T_{max}) = 0.6108 \exp\left[\frac{17.27 T_{max}}{T_{max} + 237.3}\right]$	kPa
T_{min}		°C	$e^\circ(T_{min}) = 0.6108 \exp\left[\frac{17.27 T_{min}}{T_{min} + 237.3}\right]$	kPa
saturation vapour pressure $e_s = [e^\circ(T_{max}) + e^\circ(T_{min})]/2$ Eq. 12				kPa
Actual vapour pressure: e_a				
1. e_a derived from dewpoint temperature (Eq. 14 or Table 2.3)				
T_{dew}		°C	$e_a = 0.6108 \exp\left[\frac{17.27 T_{dew}}{T_{dew} + 237.3}\right]$	kPa
OR 2. e_a derived from maximum and minimum relative humidity				
RH_{max}		%	$e^\circ(T_{min}) \frac{RH_{max}}{100}$	kPa
RH_{min}		%	$e^\circ(T_{max}) \frac{RH_{min}}{100}$	kPa
$e_a = [e^\circ(T_{min})\ RH_{max}/100 + e^\circ(T_{max})\ RH_{min}/100]/2$ Eq. 17				kPa
OR 3. e_a derived from maximum relative humidity (errors in RH_{min})				
RH_{max}		%	$e_a = e^\circ(T_{min})\ RH_{max}/100$ Eq. 18	kPa
OR 4. e_a derived from mean relative humidity (less recommended)				
RH_{mean}		%	$e_a = e_s\ (RH_{mean})/100$ Eq. 19	kPa
Vapour pressure deficit: ($e_s - e_a$)				kPa

RADIATION

Extraterrestrial Radiation (Ra)

The radiation striking a surface perpendicular to the sun's rays at the top of the earth's atmosphere, called the solar constant, is about 0.082 MJ m^{-2} min^{-1}.

The local intensity of radiation is, however, determined by the angle between the direction of the sun's rays and the normal to the surface of the atmosphere.

This angle will change during the day and will be different at different latitudes and in different seasons. The solar radiation received at the top of the earth's atmosphere on a horizontal surface is called the extraterrestrial (solar) radiation, R_a.

If the sun is directly overhead, the angle of incidence is zero and the extraterrestrial radiation is 0.0820 MJ m^{-2} min^{-1}. As seasons change, the position of the sun, the length of the day and, hence, R_a change as well.

Extraterrestrial radiation is thus a function of latitude, date and time of day. Daily values of R_a throughout the year for different latitudes are plotted in figure.

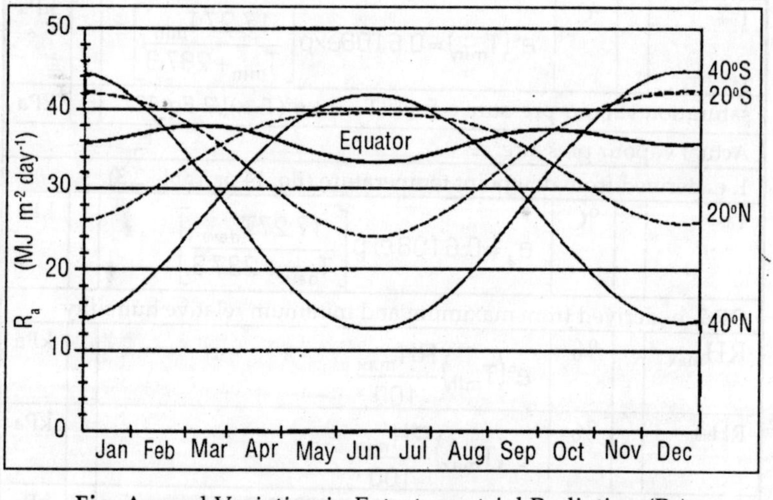

Fig. Annual Variation in Extraterrestrial Radiation (R_a) at the Equator, 20 and 40° North and South

Solar or Shortwave Radiation (Rs)

As the radiation penetrates the atmosphere, some of the radiation is scattered, reflected or absorbed by the atmospheric gases, clouds and dust. The amount of radiation reaching a horizontal plane is known as the solar radiation, R_s.

Because the sun emits energy by means of electromagnetic waves characterized by short wavelengths, solar radiation is also referred to as shortwave radiation.

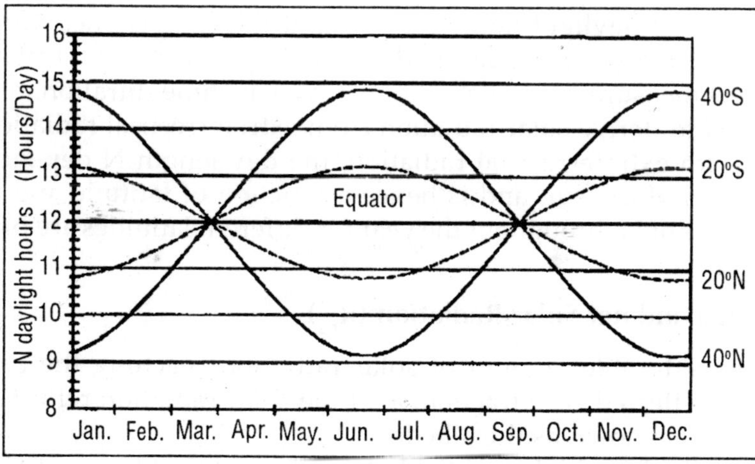

Fig. Annual Variation of the Daylight Hours (N) at the Equator, 20 and 40° North and South

For a cloudless day, R_s is roughly 75% of extraterrestrial radiation. On a cloudy day, the radiation is scattered in the atmosphere, but even with extremely dense cloud cover, about 25% of the extraterrestrial radiation may still reach the earth's surface mainly as diffuse sky radiation.

Solar radiation is also known as global radiation, meaning that it is the sum of direct shortwave radiation from the sun and diffuse sky radiation from all upward angles.

Relative Shortwave Radiation (R_s/R_{so})

The relative shortwave radiation is the ratio of the solar radiation (R_s) to the clear-sky solar radiation (R_{so}). R_s is the solar radiation that actually reaches the earth's surface in a given period, while R_{so} is the solar radiation that would reach the same surface during the same period but under cloudless conditions.

The relative shortwave radiation is a way to express the cloudiness of the atmosphere; the cloudier the sky the smaller the ratio. The ratio varies between about 0.33 (dense cloud cover) and 1 (clear sky). In the absence of a direct measurement of R_n, the relative shortwave radiation is used in the computation of the net longwave radiation.

Relative Sunshine Duration (n/N)

The relative sunshine duration is another ratio that expresses the cloudiness of the atmosphere. It is the ratio of the actual duration of sunshine, n, to the maximum possible duration of sunshine or daylight hours N. In the absence of any clouds, the actual duration of sunshine is equal to the daylight hours (n = N) and the ratio is one, while on cloudy days n and consequently the ratio may be zero. In the absence of a direct measurement of R_s, the relative sunshine duration, n/N, is often used to derive solar radiation from extraterrestrial radiation.

As with extraterrestrial radiation, the day length N depends on the position of the sun and is hence a function of latitude and date. Daily values of N throughout the year for different latitudes are plotted in figure.

Albedo (α) and net Solar Radiation (R_{ns})

A considerable amount of solar radiation reaching the earth's surface is reflected. The fraction, a, of the solar radiation reflected by the surface is known as the albedo.

The albedo is highly variable for different surfaces and for the angle of incidence or slope of the ground surface. It may be as large as 0.95 for freshly fallen snow and as small as 0.05 for a wet bare soil. A green vegetation cover has an albedo of about 0.20-0.25. For the green grass reference crop, a is assumed to have a value of 0.23.

The net solar radiation, R_{ns}, is the fraction of the solar radiation R_s that is not reflected from the surface. Its value is (1-a)R_s.

Net Longwave Radiation (R_{nl})

The solar radiation absorbed by the earth is converted to heat energy. By several processes, including emission of radiation, the earth loses this energy. The earth, which is at a much lower temperature than the sun, emits radiative energy with wavelengths longer than those from the sun. Therefore, the terrestrial radiation is referred to as longwave radiation.

The emitted longwave radiation ($R_{l,\,up}$) is absorbed by the atmosphere or is lost into space. The longwave radiation received by the atmosphere ($R_{l,\,down}$) increases its temperature and, as a consequence, the atmosphere radiates energy of its own. Part of the radiation finds it way back to the earth's surface. Consequently, the earth's surface both emits and receives longwave radiation.

The difference between outgoing and incoming longwave radiation is called the net longwave radiation, R_{nl}. As the outgoing longwave radiation is almost always greater than me incoming longwave radiation, R_{nl} represents an energy loss.

Net Radiation (R_n)

The net radiation, R_n, is the difference between incoming and outgoing radiation of both short and long wavelengths. It is the balance between the energy absorbed, reflected and emitted by the earth's surface or the difference between the incoming net shortwave (R_{ns}) and the net outgoing longwave (R_{nl}) radiation. R_n is normally positive during the daytime and negative during the nighttime. The total daily value for R_n is almost always positive over a period of 24 hours, except in extreme conditions at high latitudes.

Soil Heat Flux (G)

In making estimates of evapotranspiration, all terms of the energy balance should be considered. The soil heat flux, G, is the energy that is utilized in heating the soil. G is positive when the soil is warming and negative when the soil is cooling.

Although the soil heat flux is small compared to R_n and may often be ignored, the amount of energy gained or lost by the soil in this process should theoretically be subtracted or added to R_n when estimating evapotranspiration.

UNITS

The standard unit used in this handbook to express energy received on a unit surface per unit time is megajoules per square metre per day (MJ m^{-2} day^{-1}). In meteorological bulletins other units might be used or radiation might even be expressed in units no longer accepted as standard S. I. units, such as calories cm^{-2} day^{-1}.

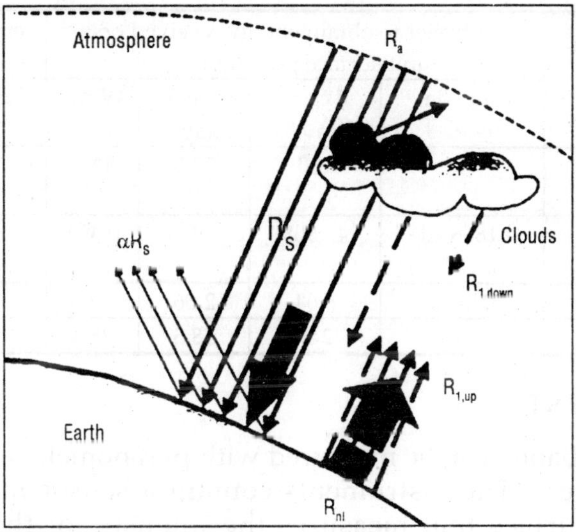

Fig. Various Components of Radiation

In the FAO Penman-Monteith equation, radiation expressed in MJ m^{-2} day^{-1} is converted to equivalent evaporation in mm day^{-1} by using

a conversion factor equal to the inverse of the latent heat heat of vaporization (1/l = 0.408):
Equivalent evaporation [mm day^{-1}] = 0.408 × Radiation [MJ m^{-2} day^{-1}]

Conversion from Energy Values to Equivalent Evaporation

The conversion from energy values to depths of water or vice versa is given by:

$$\text{Radiation [depth of Water]} = \frac{\text{Radiation [Energy/Surface]}}{\lambda P_W}$$

Where, λ latent heat of vaporization [MJ kg^{-1}]
ρ_W density of water, i.e., 1000 kg m^{-3}
[depth of water] is expressed in m
[energy/surface] is expressed in MJ m^{-2}.
By using a single value of 2.45 MJ kg^{-1} for l and multiplying the above equation by 1000 to obtain mm:

$$\text{Radiation [mm day}^{-1}] \approx \frac{\text{Radiation [MJm}^{-2}\text{ day}^{-1}]}{2.45}$$

$$= 0.408 \times \text{Radiation [MJm}^{-2}\text{day}^{-1}]$$

Common units used to express energy received on a unit surface per unit time, and conversion factors are summarized in Table.

Table. Conversion Factors for Radiation

	multiplier to obtain energy received on a unit surface per unit time				equivalent evaporation
	MJ m^{-2} day^{-1}	J cm^{-2} day^{-1}	cal cm^{-2} day^{-1}	W m^{-2}	mm day^{-1}
1 MJ m^{-2} day^{-1}	1	100	23.9	11.6	0.408
1 cal cm^{-2} day^{-1}	4.1868 10^{-2}	4.1868	1	0.485	0.0171
1 W m^{-2}	0.0864	8,64	2.06	1	0.035
1 mm day^{-1}	2.45	245	58.5	28.4	1

MEASUREMENT

Solar radiation can be measured with pyranometers, radiometers or solarimeters. The instruments contain a sensor installed on a horizontal surface that measures the intensity of the total solar radiation, i.e., both direct and diffuse radiation from cloudy conditions. The sensor is often protected and kept in a dry atmosphere by a glass dome that should be regularly wiped clean.

Net longwave and net shortwave radiation can be measured by recording the difference in output between sensors facing upward and downward. In a net radiometer, the glass domes are replaced by polyethylene domes that have a transmission range for both shortwave and longwave radiation.

Where pyranometers are not available, solar radiation is usually estimated from the duration of bright sunshine. The actual duration of sunshine, n, is measured with a Campbell-Stokes sunshine recorder. This instrument records periods of bright sunshine by using a glass globe that acts as a lens.

The sun rays are concentrated at a focal point that burns a hole in a specially treated card mounted concentrically with the sphere. The movement of the sun changes the focal point throughout the day and a trace is drawn on the card. If the sun is obscured, the trace is interrupted. The hours of bright sunshine are indicated by the lengths of the line segments.

The quantity of heat conducted into the soil, G, can be measured with systems of soil heat flux plates and thermocouples or thermisters.

Calculation Procedures

Extraterrestrial Radiation for Daily Periods (R_a)

The extraterrestrial radiation, R_a, for each day of the year and for different latitudes can be estimated from the solar constant, the solar declination and the time of the year by:

$$R_a = \frac{24(60)}{\pi} G_{sc} d_r [\omega_s \sin(\varphi)\sin(\delta) + \cos(\varphi)\cos(\delta)\sin(\omega_s)]$$

where, R_a extraterrestrial radiation [MJ m^{-2} day^{-1}],
G_{sc} solar constant = 0.0820 MJ m^{-2} min^{-1},
d_r inverse relative distance Earth-Sun ,
ω_s sunset hour angle [rad],
φ latitude [rad],
δ solar decimation (Equation 24) [rad].

R_a is expressed in the above equation in MJ m^{-2} day^{-1}. The corresponding equivalent evaporation in mm day^{-1} is obtained by multiplying R_a by 0.408. The latitude, j, expressed in radians is positive for the northern hemisphere and negative for the southern hemisphere. The conversion from decimal degrees to radians is given by:

$$[\text{Radians}] = \frac{\pi}{180}[\text{decimal deg rees}]$$

Example. Conversion of latitude in degrees and minutes to radians

Express the latitudes of Bangkok (Thailand) at 13°44'N and Rio de Janeiro (Brazil) at 22°54'S in radians.		
Latitude	Bangkok (northern hemisphere)	Rio de Janeiro (southern hemisphere)
degrees & minutes	13°44'N	22°54'S
decimal degrees	13 + 44/60=13.73	(-22) + (-54/60) = - 22.90
radians	(π /180) 13.73 = + 0.240	(π /180) (-22.90) = - 0.400
The latitudes of Bangkok and Rio de Janeiro are respectively +0.240 and -0.400 radians.		

The inverse relative distance Earth-Sun, d_r, and the solar declination, δ, are given by:

$$d_r = 1 + 0.033\cos\left(\frac{2\pi}{365}J\right)$$

$$\delta = 0.409\sin\left(\frac{2\pi}{365}J - 1.39\right)$$

where J is the number of the day in the year between 1 (1 January) and 365 or 366 (31 December). Values for J for all days of the year and an equation for estimating J. The sunset hour angle, ω_s, is given by:

$$\omega_s = \arccos\,[-\tan(\varphi)\tan(\delta)]$$

As the arccos function is not available in all computer languages, the sunset hour angle can also be computed using the arctan function:

$$\omega_s = \frac{\pi}{2} - \arctan\left[\frac{-\tan(\varphi)\tan(\delta)}{X^{0.5}}\right]$$

where

$$X = 1 - [\tan(\varphi)]^2\,[\tan(\delta)]^2$$

and $X = 0.00001$ if $X \leq 0$

These values represent R_a on the 15th day of each month. These values deviate from values that are averaged over each day of the month by less than 1% for all latitudes during non-frozen periods and are included for simplicity of calculation.

These values deviate slightly from the values in the Smithsonian Tables.

For the winter months in latitudes greater than 55° (N or S), the equations for R_a have limited validity. Reference should be made to the Smithsonian Tables to assess possible deviations.

Example. Determination of Extraterrestrial Radiation

Determine the extraterrestrial radiation (R_a) for 3 September at 20°S.			
From Eq. 22	20°S or $\varphi = (\pi/180)(-20)$ = (the value is negative for the southern hemisphere)	-0.35	rad
From Table 2.5:	The number of day in the year, J =	246	days
From Eq. 23	$d_r = 1 + 0.033\cos(2\pi(246)/365)$ =	0.985	rad
From Eq. 24	$\delta = 0.409\sin(2\pi(246)/365 - 1.39)$ =	0.120	rad
From Eq. 25:	$\omega_s = \arccos[-\tan(-0.35)\tan(0.120)]$ =	1.527	rad
Then:	$\sin(\varphi)\sin(\delta)$ =	-0.041	-
and:	$\cos(\varphi)\cos(\delta)$ =	0.933	-
From Eq. 21	$R_a = 24(60)/\pi\,(0.0820)(0.985)[1.527(-0.041) + 0.933\sin(1.527)]$ =	32.2	MJ m^{-2} d^{-1}
From Eq. 20	expressed as equivalent evaporation = 0.408 (32.2) =	13.1	mm/day
The extraterrestrial radiation is 32.2 MJ m^{-2} day^{-1}.			

Extraterrestrial Radiation for Hourly or Shorter Periods (R_a)

For hourly or shorter periods the solar time angle at the beginning and end of the period should be considered when calculating R_a:

$$R_a = \frac{12(60)}{\pi} G_{sc} d_r [(\omega_2 - \omega_1)\sin(\varphi)\sin(\delta) + \cos(\varphi)\cos(\delta)(\sin(\omega_2) - \sin(\omega_1))]$$

where, R_a extraterrestrial radiation in the hour (or shorter) period [MJ m^{-2} $hour^{-1}$],

G_{sc} solar constant = 0.0820 MJ m^{-2} min^{-1},

d_r inverse relative distance Earth-Sun

δ solar declination [rad],

φ latitude [rad],

ω_1 solar time angle at beginning of period [rad],

ω_2 solar time angle at end of period [rad]

The solar time angles at the beginning and end of the period are given by:

$$\omega_1 = \omega - \frac{\pi t_1}{24}$$

$$\omega_2 = \omega + \frac{\pi t_1}{24}$$

where, ω solar time angle at midpoint of hourly or shorter period [rad],

t_1 length of the calculation period [hour]: i.e., 1 for hourly period or 0.5 for a 30-minute period.

The solar time angle at midpoint of the period is:

$$\omega = \frac{\pi}{12}[(t + 0.06667(L_z - L_m) + S_c) - 12]$$

where, t standard clock time at the midpoint of the period [hour]. For example for a period between 14.00 and 15.00 hours, t = 14.5,

L_z longitude of the centre of the local time zone [degrees west of Greenwich]. For example, L_z = 75, 90, 105 and 120° for the Eastern, Central, Rocky Mountain and Pacific time zones (United States) and L_z = 0° for Greenwich, 330° for Cairo (Egypt), and 255° for Bangkok (Thailand),

L_m longitude of the measurement site [degrees west of Greenwich],

S_c seasonal correction for solar time [hour].

Of course, $\omega < -\omega_s$ or $\omega > \omega_s$ from Equation 31 indicates that the sun is below the horizon so that, by definition, R_a is zero.

The seasonal correction for solar time is:

$$S_c = 0.1645 \sin(2b) - 0.1255 \cos(b) - 0.025 \sin(b)$$

$$b = \frac{2\pi(J - 81)}{364}$$

where J is the number of the day in the year.

Daylight Hours (N)

The daylight hours, N, are given by:

$$N = \frac{24}{\pi}\omega_s$$

where ω_s is the sunset hour angle in radians given by Equation. Mean values for N (15th day of each month) for different latitudes.

Example. Determination of daylight hours.

Determine the daylight hours (N) for 3 September at 20°S.			
From Example 8:	ω_s = arccos[-tan(-0.35)tan(0.120)] =	1.527	rad
From Eq. 34:	N= 24/π (1.527) =	11.7	hour
The number of daylight hours is 11.7 hours.			

Solar Radiation (R_s)

If the solar radiation, R_s, is not measured, it can be calculated with the Angstrom formula which relates solar radiation to extraterrestrial radiation and relative sunshine duration:

$$R_s = \left(a_s + b_s \frac{n}{N}\right) R_a$$

where, R_s solar or shortwave radiation [MJ m^{-2} day^{-1}],

n actual duration of sunshine [hour],

N maximum possible duration of sunshine or daylight hours [hour], n/N relative sunshine duration [-],

R_a extraterrestrial radiation [MJ m^{-2} day^{-1}],

a_s regression constant, expressing the fraction of extraterrestrial radiation reaching the earth on overcast days (n = 0),

a_s+b_s fraction of extraterrestrial radiation reaching the earth on clear days (n = N).

R_s is expressed in the above equation in MJ m^{-2} day^{-1}. The corresponding equivalent evaporation in mm day^{-1} is obtained by multiplying R_s by 0.408.

Depending on atmospheric conditions (humidity, dust) and solar declination (latitude and month), the Angstrom values a_s and b_s will vary.

Where no actual solar radiation data are available and no calibration has been carried out for improved a_s and b_s parameters, the values $a_s = 0.25$ and $b_s = 0.50$ are recommended.

The extraterrestrial radiation, R_a, and the daylight hours or maximum possible duration of sunshine, N, are given by Equations. The actual duration of sunshine, n, is recorded with a Campbell Stokes sunshine recorder.

Clear-Sky Solar Radiation (R_{so})

The calculation of the clear-sky radiation, R_{so}, when n = N, is required for computing net longwave radiation.

- For near sea level or when calibrated values for a_s and b_s are available:

$$R_{so} = (a_s+b_s)R_a$$

where, R_{so} clear-sky solar radiation [MJ m^{-2} day^{-1}],

a_s+b_s fraction of extraterrestrial radiation reaching the earth on clear-sky days (n = N).

- When calibrated values for a_s and b_s are not available:

$$R_{so} = (0.75 + 2\ 10^{-5}z)R_a$$

where, z station elevation above sea level [m].

Net Solar or Net Shortwave Radiation (R_{ns})

The net shortwave radiation resulting from the balance between incoming and reflected solar radiation is given by:

$$R_{ns} = (1-\alpha)R_s$$

where, R_{ns} net solar or shortwave radiation [MJ m^{-2} day^{-1}],

α albedo or canopy reflection coefficient, which is 0.23 for the hypothetical grass reference crop [dimensionless],

R_s the incoming solar radiation [MJ m^{-2} day^{-1}].

R_{ns} is expressed in the above equation in MJ m^{-2} day^{-1}.

Net Longwave Radiation (R_{nl})

The rate of longwave energy emission is proportional to the absolute temperature of the surface raised to the fourth power. This relation is expressed quantitatively by the Stefan-Boltzmann law. The net energy flux leaving the earth's surface is, however, less than that emitted and given by the Stefan-Boltzmann law due to the absorption and downward radiation from the sky.

Water vapour, clouds, carbon dioxide and dust are absorbers and emitters of longwave radiation. Their concentrations should be known when assessing the net outgoing flux. As humidity and cloudiness play an important role, the Stefan-Boltzmann law is corrected by these two factors when estimating - the net outgoing flux of longwave radiation. It is thereby assumed that the concentrations of the other absorbers are constant:

$$R_{nl} = \sigma\left[\frac{T_{max\,k}{}^4 + T_{min\,k}{}^4}{2}\right]\left(0.34 - 0.14\sqrt{e_a}\right)\left(1.35\frac{R_s}{R_{so}} - 0.35\right)$$

where, R_{nl} net outgoing longwave radiation [MJ m^{-2} day^{-1}],
σ Stefan-Boltzmann constant [4.903 10^{-9} MJ K^{-4} m^{-2} day^{-1}],
$T_{max,K}$ maximum absolute temperature during the 24-hour period [K = °C + 273.16],
$T_{min,K}$ minimum absolute temperature during the 24-hour period [K = °C + 273.16],
e_a actual vapour pressure [kPa],
R_s/R_{so} relative shortwave radiation (limited to $\leq$ 1.0),
R_s measured or calculated. solar radiation [MJ m^{-2} day^{-1}],
R_{so} calculated clear-sky radiation [MJ m^{-2} day^{-1}].

An average of the maximum air temperature to the fourth power and the minimum air temperature to the fourth power is commonly used in the Stefan-Boltzmann equation for 24-hour time steps.

The term $(0.34 - 0.14\sqrt{e_a})$ expresses the correction for air humidity, and will be smaller if the humidity increases. The effect of cloudiness is expressed by $(1.35\ R_s/R_{so} - 0.35)$.

The term becomes smaller if the cloudiness increases and hence R_s decreases.

The smaller the correction terms, the smaller the net outgoing flux of longwave radiation. Note that the R_s/R_{so} term in Equation must be limited so that $R_s/R_{so} \leq 1.0$.

Where measurements of incoming and outgoing short and longwave radiation during bright sunny and overcast hours are available, calibration of the coefficients in Equation can be carried out.

Net Radiation (R_n)

The net radiation (R_n) is the difference between the incoming net shortwave radiation (R_{ns}) and the outgoing net longwave radiation (R_{nl}):

$$R_n = R_{ns} - R_{nl}$$

Example. Determination of net radiation.

Determine the net radiation in Rio de Janeiro in May with the data from previous examples.			
From Example:	R_s =	14.5	MJ m^{-2} day^{-1}
From Equation:	R_{ns} = (1 - 0.23) R_s =	11.1	MJ m^{-2} day^{-1}
From Example:	R_{nl} =	3.5	MJ m^{-2} day^{-1}
From Equation:	R_n = 11.1-3.5 =	7.6	MJ m^{-2} day^{-1}
From Equation:	expressed as equivalent evaporation = 0.408 (7.7) =	3.1	mm/day
The net radiation is 7.6 MJ m^{-2} day^{-1}.			

SOIL HEAT FLUX (G)

Complex models are available to describe soil heat flux. Because soil heat flux is small compared to R_n, particularly when the surface is covered by vegetation and calculation time steps are 24 hours or longer, a simple calculation procedure is presented here for long time steps, based on the idea that the soil temperature follows air temperature:

$$G = c_s \frac{T_i - T_{i-1}}{\Delta t} \Delta z$$

where, G soil heat flux [MJ m^{-2} day^{-1}],

c_s soil heat capacity [MJ m^{-3} °C^{-1}],

T_i air temperature at time i [°C],

T_{i-1} air temperature at time i – 1 [°C],

Δt length of time interval [day],

Δz effective soil depth [m].

As the soil temperature lags air temperature, the average temperature for a period should be considered when assessing the daily soil heat flux, i.e., Δt should exceed one day. The depth of penetration of the temperature wave is determined by the length of the time interval. The effective soil depth, Δz, is only 0.10 – 0.20 m for a time interval of one or a few days but might be 2 m or more for monthly periods. The soil heat capacity is related to its mineral composition and water content.

- For day and ten-day periods:

As the magnitude of the day or ten-day soil heat flux beneath the grass reference surface is relatively small, it may be ignored and thus:

$$G_{day} \approx 0$$

- For monthly periods:

When assuming a constant soil heat capacity of 2.1 MJ m^{-3} °C^{-1}

and an appropriate soil depth, Equation 41 can be used to derive G for monthly periods:

$$G_{month,\ i} = 0.07\ (T_{month,\ i+1} - T_{month,\ i-1})$$

or, if $T_{month,\ i+1}$ is unknown:

$$G_{month,\ i} = 0.14\ (T_{month,\ i} - T_{month,\ i-1})$$

where. $T_{month,\ i}$ mean air temperature of month i [°C],
$T_{month,\ i-1}$ mean air temperature of previous month [°C],
$T_{month,\ i+1}$ mean air temperature of next month [°C].

- For hourly or shorter periods:

For hourly (or shorter) calculations, G beneath a dense cover of grass does not correlate well with air temperature. Hourly G can be approximated during daylight periods as:

$G_{hr} = 0.1\ R_n$ and during nighttime periods as:

$$G_{hr} = 0.5\ R_n$$

Where the soil is warming, the soil heat flux G is positive. The amount of energy required for this process is subtracted from R_n when estimating evapotranspiration.

Example. Determination of soil heat flux for monthly periods

Determine the soil heat flux in April in Algiers (Algeria) when the soil is warming. The mean monthly temperatures of March, April and May are 14.1, 16.1, and 18.8°C respectively.			
From Equation \	for the month of April: $G_{month} = 0.07\ (18.8 - 14.1) =$	0.33	MJ m^{-2} day^{-1}
From Equation	expressed as equivalent evaporation = 0.408(0.33) =	0.13	mm/day
The soil heat flux is 0.33 MJ m^{-2} day^{-1}.			

WIND SPEED

Measurement

Wind is characterized by its direction and velocity. Wind direction refers to the direction from which the wind is blowing. For the computation of evapotranspiration, wind speed is the relevant variable.

As wind speed at a given location varies with time, it is necessary to express it as an average over a given time interval. Wind speed is given in metres per second (m s^{-1}) or kilometres per day (km day^{-1}).

Wind speed is measured with anemometers. The anemometers commonly used in weather stations are composed of cups or propellers which are turned by the force of the wind. By counting the number of revolutions over a given time period, the average wind speed over the measuring period is computed.

Wind Profile Relationship

Wind speeds measured at different heights above the soil surface are different. Surface friction tends to slow down wind passing over it. Wind speed is slowest at the surface and increases with height. For this reason anemometers are placed at a chosen standard height, i.e., 10 m in meteorology and 2 or 3 m in agrometeorology.

For the calculation of evapotranspiration, wind speed measured at 2 m above the surface is required. To adjust wind speed data obtained from instruments placed at elevations other than the standard height of 2m, a logarithmic wind speed profile may be used for measurements above a short grassed surface:

$$U_2 = U_Z \frac{4.87}{\ln(67.8z - 5.42)}$$

where, u_2 wind speed at 2 m above ground surface [m s^{-1}],
u_z measured wind speed at z m above ground surface [m s^{-1}],
z height of measurement above ground surface [m].

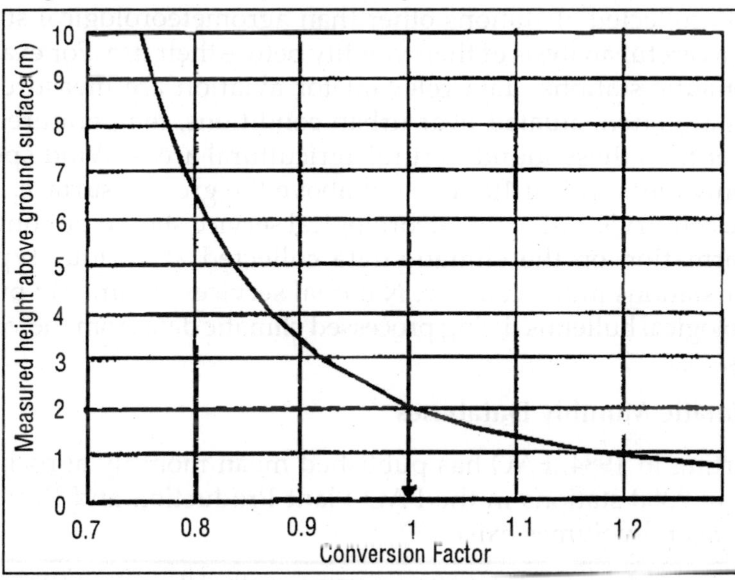

Fig. Conversion Factor to Convert Wind Speed Measured at a Certain Height above Ground Level to Wind Speed at the Standard Height (2 m)

Example. Adjusting wind speed data to standard height

Determine the wind speed at the standard height of 2 m, from a measured wind speed of 3.2 m/s at 10 m above the soil surface.			
For:	u_z =	3.2	m/s
And:	z =	10	m
Then:	Conversion factor = 4.87/ln (67.8 (10) - 5.42) =	0.75	-
From Equation:	u_2 = 3.2 (0.75) =	2.4	m/s
The wind speed at 2 m above the soil surface is 2.4 m/s.			

CLIMATIC DATA ACQUISITION

Weather Stations

Meteorological data are recorded at various types of weather stations. Agrometeorological stations are sited in cropped areas where instruments are exposed to atmospheric conditions similar to those for the surrounding crops. In these stations, air temperature and humidity, wind speed and sunshine duration are typically measured at 2 m above an extensive surface of grass or short crop.

Where needed and feasible, the cover of the station is irrigated. Guidelines for the establishment and maintenance of agrometeorological stations are given in the FAO Irrigation and Drainage. This handbook also describes the different types of instruments, their installation and reliability.

Data collected at stations other than agrometeorological stations require a careful analysis of their validity before their use. For example, in aeronautic stations, data relevant for aviation are measured. As airports are often situated near urban conditions, temperatures may be higher than those found in rural agricultural areas. Wind speed is commonly measured at 10 m height above the ground surface.

The country's national meteorological service should be contacted for information on the climatic data collected at various types of weather stations in the country. National services commonly publish meteorological bulletins listing processed climatic data from the various stations.

Agroclimatic Monthly Databases

Starting in 1984, FAO has published mean monthly agroclimatic data from 2300 stations in the FAO Plant Production and Protection Series. Several volumes exist:

No. 22:	Volume 1: data for Africa, countries north of the equator (1984),
	Volume 2: data for Africa, countries south of the equator (1984);
No. 24:	Agroclimatic data for Latin America and the Caribbean (1985);
No. 25:	Volume 1: Agroclimatic data for Asia (A-J) (1987),
	Volume 2: Agroclimatic data for Asia (K-Z) (1987).

CLIMWAT for CROPWAT contains monthly data from 3 262 climatic stations contained on five separate diskettes. The stations are grouped by country and by continent.

Monthly averages of maximum and minimum temperatures, mean relative humidity, wind speed, sunshine hours, radiation data as well as rainfall and ET_0 calculated with the FAO Penman-Monteith method are listed on the diskettes for mean long-term conditions. FAOCLIM provides a user friendly interface on compact disc to the agroclimatic database of the Agrometeorology Group in FAO.

The data presented are an extension of the previously published FAO Plant Production and Protection Series and the number of stations has been increased from 2300 to about 19000, with an improved world wide coverage. However, values for all principal weather parameters are not available for all stations. Many contain air temperature and precipitation only.

These databases can be consulted in order to verify the consistency of the actual database or to estimate missing climatic parameters. However, they should only be used for preliminary studies as they contain mean monthly data only. FAOCLIM provides monthly time series for only a few stations. The information in these databases should never replace actual data.

Other electronic databases for portions of the globe have been published by the International Water Management Institute (IWMI). These databases include daily and monthly air temperature, precipitation and ET_o predicted using the Hargreaves ET_o equation that is based on differences between daily maximum and minimum air temperature.

Estimating Missing Climatic Data

The assessment of the reference evapotranspiration ET_o with the Penman-Monteith method. The calculation requires mean daily, ten-day or monthly maximum and minimum air temperature (T_{max} and T_{min}), actual vapour pressure (e_a), net radiation (R_n) and wind speed measured at 2 m (u_2).

If some of the required weather data are missing or cannot be calculated, it is strongly recommended that the user estimate the missing climatic data with one of the following procedures and use the FAO Penman-Monteith method for the calculation of ET_o. The use of an alternative ET_o calculation procedure, requiring only limited meteorological parameters, is less recommended. Procedures to estimate missing humidity, radiation and wind speed data are given in this section.

Estimating Missing Humidity Data

Where humidity data are lacking or are of questionable quality, an estimate of actual vapour pressure, e_a, can be obtained by assuming that dewpoint temperature (T_{dew}) is near the daily minimum temperature (T_{min}). This statement implicitly assumes that at sunrise, when the air temperature is close to T_{min}, that the air is nearly saturated with water vapour and the relative humidity is nearly 100%. If T_{min} is used to represent T_{dew} then:

$$e_a = e^o(T_{min}) = 0.611 \exp\left[\frac{17.27 T_{min}}{T_{min} + 237.3}\right]$$

The relationship $T_{dew} \approx T_{min}$ holds for locations where the cover crop of the station is well watered. However, particularly for arid regions, the air might not be saturated when its temperature is at its minimum. Hence, T_{min} might be greater than T_{dew} and a further calibration may be required to estimate dewpoint temperatures. In these situations, "T_{min}" in the above equation may be better approximated by subtracting 2-3 °C from T_{min}.

In humid and subhumid climates, T_{min} and T_{dew} measured in early morning may be less than T_{dew} measured during the daytime because of condensation of dew during the night. After sunrise, evaporation of the dew will once again humidify the air and will increase the value measured for T_{dew} during the daytime. However, it is standard practice in 24-hour calculations of ET_o to use T_{dew} measured or calculated during early morning.

The estimate for e_a from T_{min} should be checked. When the prediction by Equation 48 is validated for a region, it can be used for daily estimates of e_a.

Estimating Missing Radiation Data

Net radiation measuring devices, requiring professional control, have rarely been installed in agrometeorological stations. In the absence of a direct measurement, longwave and net radiation can be derived from more commonly observed weather parameters, i.e., solar radiation or sunshine hours, air temperature and vapour pressure. Where solar radiation is not measured, it can perhaps be estimated from measured hours of bright sunshine.

However, where daily sunshine hours (n) are not available, solar radiation data cannot be computed with the calculation procedures previously presented. This section presents various methods to estimate solar radiation data with an alternative methodology.

Solar Radiation Data from a Nearby Weather Station

This method relies on the fact that for the same month and often for the same day, the variables affecting incoming solar radiation, R_s, and sunshine duration, n, are similar throughout a given region. This implies that:

- The size of the region is small.
- The air masses governing rainfall and cloudiness are nearly identical within parts of the region.
- The physiography of the region is almost homogenous.

Differences in relief should be negligible as they strongly influence the movement of air masses. Under such conditions, radiation data observed at nearby stations can be used.

Caution should be used when applying this method to mountainous and coastal areas where differences in exposure and altitude could be important or where rainfall is variable due to convective conditions. Moreover, data from a station located nearby but situated on the other side of a mountain may not be transferable as conditions governing radiation are different. The user should observe climatic conditions in both locations and obtain information from local persons concerning general differences in cloud cover and type.

Where the north-south distance to a weather station within the same homogeneous region exceeds 50 km so that the value for R_a changes, the R_s measurement should be adjusted using the ratio of the solar to extraterrestrial radiation, R_s/R_a:

$$R_s = \frac{R_{s.reg}}{R_{a.reg}} R_a$$

where

$R_{s,\ reg}$ solar radiation at the regional location [MJ m^{-2} day^{-1}],

$R_{a,\ reg}$ extraterrestrial radiation at the regional location [MJ m^{-2} day^{-1}]

Once the solar radiation has been derived from the radiation data of a nearby station, the net longwave radiation (Equation 39) and the net radiation can be calculated.

The estimation method of Equation 49 is recommended for monthly calculations of ET_o. If using the method for daily estimates of ET_o, a more careful analysis of weather data in the importing and exporting meteorological stations has to be performed to verify whether both stations are in the same homogeneous climatic region and are close enough to experience similar conditions within the same day.

The analysis should include the comparison of daily weather data from both stations, particularly the maximum and minimum air temperature and humidity. In fact, similar cloudiness and sunshine durations are related to similarities in temperature and humidity trends.

Generally, daily calculations of ET_o with estimated radiation data are justified when utilized as a sum or an average over a several-day period. This is the case for the computation of the mean evapotranspiration demand between successive irrigations or when planning irrigation schedules.

Under these conditions, the relative error for one day often counterbalances the error for another day of the averaging period. Daily estimates should not be utilized as true daily estimates but only in averages over the period under consideration.

Solar Radiation Data Derived from Air Temperature Differences

The difference between the maximum and minimum air temperature is related to the degree of cloud cover in a location. Clear-sky conditions result in high temperatures during the day (T_{max},) because the atmosphere is transparent to the incoming solar radiation and in low temperatures during the night (T_{min}) because less outgoing longwave radiation is absorbed by the atmosphere.

On the other hand, in overcast conditions, T_{max} is relatively smaller because a significant part of the incoming solar radiation never reaches the earth's surface and is absorbed and reflected by the clouds. Similarly, T_{min} will be relatively higher as the cloud cover acts as a blanket and decreases the net outgoing longwave radiation.

Therefore, the difference between the maximum and minimum air temperature (T_{max} - T_{min}) can be used as an indicator of the fraction of extraterrestrial radiation that reaches the earth's surface. This principle has been utilized by Hargreaves and Samani to develop estimates of ET_o using only air temperature data.

The Hargreaves' radiation formula, adjusted and validated at several weather stations in a variety of climate conditions, becomes:

$$R_s = K_{Rs}\sqrt{(T_{max} - T_{min})}R_a$$

Where, R_a extraterrestrial radiation [MJ m^{-2} d^{-1}],

T_{max} maximum air temperature [°C],

T_{min} minimum air temperature [°C],

k_{Rs} adjustment coefficient (0.16.. 0.19) [°$C^{-0.5}$].

The square root of the temperature difference is closely related to the existing daily solar radiation in a given location. The adjustment coefficient k_{Rs} is empirical and differs for 'interior' or 'coastal' regions:

- For 'interior' locations, where land mass dominates and air masses are not strongly influenced by a large water body, $k_{Rs} \cong 0.16$;
- For 'coastal' locations, situated on or adjacent to the coast of a large land mass and where air masses are influenced by a nearby water body, $k_{Rs} \cong 0.19$.

The relationship between R_s/R_a and the temperature difference is plotted for interior and coastal locations. The fraction of extraterrestrial radiation that reaches the earth's surface, R_s/R_a, ranges from about 0.25 on a day with dense cloud cover to about 0.75 on a cloudless day with clear sky.

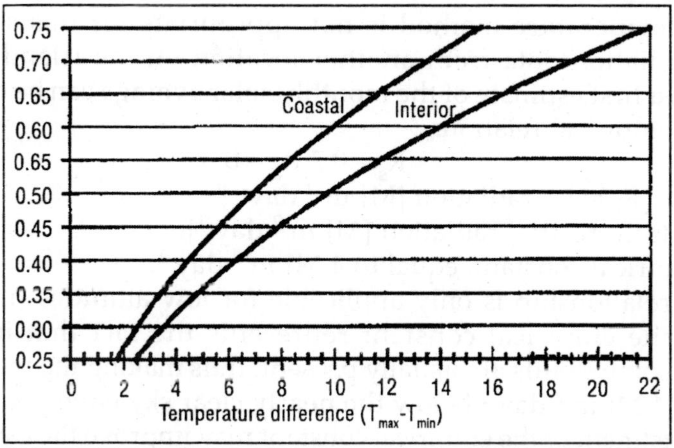

Fig. Relationship Between the Fraction of Extraterrestrial Radiation that Reaches the Earth's Surface, R_s/R_a, and the Air Temperature Difference $T_{max} - T_{min}$ for Interior ($k_{Rs} = 0.16$) and Coastal ($k_{Rs} = 0.19$) Regions

The temperature difference method is recommended for locations where it is not appropriate to import radiation data from a regional station, either because homogeneous climate conditions do not occur, or because data for the region are lacking. For island conditions, the methodology of Equation 50 is not appropriate due to moderating effects of the surrounding water body. Caution is required when daily computations of ET_0 are needed. The advice given for Equation 49 fully applies. It is recommended that daily estimates of ET_0 that are based on estimated R_s be summed or averaged over a several-day period, such as a week, decade or month to reduce prediction error.

Example Determination of solar radiation from temperature data.

Determine the solar radiation from the temperature data of July in Lyon (France) at a latitude of 45°43'N and at 200 m above sea level. In July, the mean monthly maximum and minimum air temperatures are 26.6 and 14.8°C respectively.			
	latitude = 45°43' = +45.72" decimal degrees =	0.80	radian
From Table:	The day of the year for 15 July is	196	-
From Equation and Table:	R_a =	40.6	MJ m^{-2} day^{-1}
From Equation (same latitude):	$R_s = 0.16 [\sqrt{(26.6-14.8)}] R_a$ = 0.55 (40.6) =	22.3	MJ m^{-2} day^{-1}
From Equation (same latitude):	equivalent evaporation = 0.408 (22.3) =	9.1	mm/day
In July, the estimated solar radiation, R_s, is 22.3 MJ m^{-2} day^{-1}			

Empirical Methodology for Island Locations

For island locations, where the land mass has a width perpendicular to the coastline of 20 km or less, the air masses influencing the atmospheric conditions are dominated by the adjacent water body in all directions.

The temperature method is not appropriate for this situation. Where radiation data from another location on the island are not available, a first estimate of the monthly solar average can be obtained from the empirical relation:

$$R_s = 0.7\ R_a - b$$

where, R_s solar radiation [MJ m^{-2} day^{-1}].

R_a extraterrestrial radiation [MJ m^{-2} day^{-1}].

b empirical constant, equal to 4 MJ m^{-2} day^{-1}.

This relationship is only applicable for low altitudes (from 0 to 100 m). The empirical constant represents the fact that in island locations some clouds are usually present, thus making the mean solar radiation 4 MJ m^{-2} day^{-1} below the nearly clear sky envelope (0.7 R_a). Local adjustment of the empirical constant may improve the estimation.

The method is only appropriate for monthly calculations. The constant relation between R_s and R_a does not yield accurate daily estimates.

MISSING WIND SPEED DATA

Wind Speed Data from a nearby Weather Station

Importing wind speed data from a nearby station, as for radiation data, relies on the fact that the air flow above a 'homogeneous' region may have relatively large variations through the course of a day but small variations when referring to longer periods or the total for the day. Data from a nearby station may be imported where air masses are of the same origin or where the same fronts govern air flows in the region and where the relief is similar.

When importing wind speed data from another station, the regional climate, trends in variation of other meteorological parameters and relief should be compared. Strong winds are often associated with low relative humidity and light winds are common with high relative humidity.

Thus, trends in variation of daily maximum and minimum relative humidities should be similar in both locations. In mountainous areas, data should not necessarily be imported from the nearest station but from nearby stations with similar elevation and exposure to the dominant winds. The paired stations may even vary from one season to another, depending on the dominant winds. Imported wind speed data can be used when making monthly estimates of evapotranspiration. Daily calculations are justified when utilized as a sum or average over a several-day period, such as a week or decade.

Empirical Estimates of Monthly Wind Speed

As the variation in wind speed average over monthly periods is relatively small and fluctuates around average values, monthly values of wind speed may be estimated. The 'average' wind speed estimates may be selected from information available for the regional climate, but should take seasonal changes into account.

Table. General Classes of Monthly wind Speed Data

Description	Mean Monthly Wind Speed at 2 m
light wind	...≤ 1.0 m/s
light to moderate wind	1 - 3 m/s
moderate to strong wind	3 - 5 m/s
strong wind	... ≥ 5.0 m/s

Where no wind data are available within the region, a value of 2 m/s can be used as a temporary estimate. This value is the average over 2000 weather stations around the globe.

In general, wind speed at 2 m, u_2, should be limited to about $u_2 \geq 0.5$ m/s when used in the ET_0 equation (Equation 6). This is necessary to account for the effects of boundary layer instability and buoyancy of air in promoting exchange of vapour at the surface when air is calm. This effect occurs when the wind speed is small and buoyancy of warm air induces air exchange at the surface. Limiting $u_2 \geq 0.5$ m/s in the ET_0 equation improves the estimation accuracy under the conditions of very low wind speed.

MINIMUM DATA REQUIREMENTS

This section has shown how solar radiation, vapour pressure and wind data can be estimated when missing. Many of the suggested procedures rely upon maximum and minimum air temperature measurements.

Unfortunately, there is no dependable way to estimate air temperature when it is missing. Therefore it is suggested that maximum and minimum daily air temperature data are the minimum data requirements necessary to apply the FAO Penman-Monteith method.

An Alternative Equation for ET_0 when weather Data are Missing

When solar radiation data, relative humidity data and/or wind speed data are missing, they should be estimated using the procedures presented in this section. As an alternative, ET_o can be estimated using the Hargreaves ET_o equation where:

$$ET_o = 0.0023(T_{mean} + 17.8)(T_{max} - T_{min})^{0.5} R_a$$

where all parameters have been previously defined. Units for both ET_o and R_a in Equation are mm day^{-1}. Equation should be verified in each new region by comparing with estimates by the FAO Penman-Monteith equation at weather stations where solar radiation, air temperature, humidity, and wind speed are measured.

If necessary, Equation can be calibrated on a monthly or annual basis by determining empirical coefficients where ET_o = a + b ET_o Equation , subscript refers to ET_o predicted using Equation. The coefficients a and b can be determined by regression analyses or by visual fitting.

In general, estimating solar radiation, vapour pressure and wind speed as described in Equations and Table and then utilizing these estimates in Equation (the FAO Penman-Monteith equation) will provide somewhat more accurate estimates as compared to estimating ET_o directly using Equation. This is due to the ability of the estimation equations to incorporate general climatic characteristics such as high or low wind speed or high or low relative humidity into the ET_o estimate made using Equation.

Equation has a tendency to underpredict under high wind conditions (u_2 > 3 m/s) and to overpredict under conditions of high relative humidity.

Chapter 4

Determination of ET_o

This chapter demonstrates how the crop reference evapotranspiration (ET_o) is determined either from meteorological data or from pan evaporation.

The FAO Penman-Monteith method is maintained as the sole standard method for the computation of ET_o from meteorological data. This chapter presents guidelines to calculate ET_o with different time steps, ranging from hours to months, and with missing climatic data. The ET_o calculation can be done by hand with the help of a calculation sheet, or by means of a computer.

ET_o can also be estimated from the evaporation loss from a water surface. The procedure to obtain ET_o from pan evaporation and the coefficients for different types of pans are presented in this chapter.

Penman-Monteith Equation

From the original Penman-Monteith equation and the equations of the aerodynamic and canopy resistance, the FAO Penman-Monteith equation has been derived:

$$ET_o = \frac{0.408\Delta(R_n - G) + \gamma \frac{900}{T+273} U_2(e_s - e_a)}{\Delta + \gamma(1 + 0.34U_2)}$$

where, ET_o reference evapotranspiration [mm day^{-1}],
R_n net radiation at the crop surface [MJ m^{-2} day^{-1}],
G soil heat flux density [MJ m^{-2} day^{-1}],
T air temperature at 2 m height [°C],
u_2 wind speed at 2 m height [m s^{-1}],
e_s saturation vapour pressure [kPa],
e_a actual vapour pressure [kPa],
e_s - e_a saturation vapour pressure deficit [kPa],
Δ slope vapour pressure curve [kPa °C^{-1}],
γ psychrometric constant [kPa °C^{-1}].

The FAO Penman-Monteith equation determines the evapotranspiration from the hypothetical grass reference surface and provides a standard to which evapotranspiration in different periods of the year or in other regions can be compared and to which the evapotranspiration from other crops can be related.

CALCULATION PROCEDURE

Calculation Sheet

ET_o can be estimated by means of the calculation sheet. The calculation procedure consists of the following steps:

- Derivation of some climatic parameters from the daily maximum (T_{max}) and minimum (T_{min}) air temperature, altitude (z) and mean wind speed (u_2).
- Calculation of the vapour pressure deficit ($e_s - e_a$). The saturation vapour pressure (e_s) is derived from T_{max} and T_{min}, while the actual vapour pressure (e_a) can be derived from the dewpoint temperature (T_{dew}), from maximum (RH_{max}) and minimum (RH_{min}) relative humidity, from the maximum (RH_{max}), or from mean relative humidity (RH_{mean}).
- Determination of the net radiation (R_n) as the difference between the net shortwave radiation (R_{ns}) and the net longwave radiation (R_{nl}). In the calculation sheet, the effect of soil heat flux (G) is ignored for daily calculations as the magnitude of the flux in this case is relatively small. The net radiation, expressed in MJ m^{-2} day^{-1}, is converted to mm/day (equivalent evaporation) in the FAO Penman-Monteith equation by using 0.408 as the conversion factor within the equation.
- ET_o is obtained by combining the results of the previous steps.

Computerized Calculations

Calculations of the reference crop evapotranspiration ET_o are often computerized. The calculation procedures of all data required for the calculation of ET_o by means of the FAO Penman-Monteith equation. Typical sequences in which the calculations can be executed are given in the calculation sheets.

Many software packages already use the FAO Penman-Monteith equation to assess the reference evapotranspiration.

ET_o Calculated with Different Time Steps

The selection of the time step with which ET_o is calculated depends on the purpose of the calculation, the accuracy required and the time step of the climatic data available.

MONTHLY REFERENCE EVAPOTRANSPIRATION PENMAN MONTEITH							
Mateostation: CABINDA				Country: Angola			
Altitude: 20 m.				Coordinates: -5.33		South	12.11 East
Month	MinTemp	MaxTemp	Humid.	Wind	Sunshine	Radiation	ETo-PenMon
	°C	°C	%	km/day	Hours	MJ/m²/day	mm/day
January	22.8	29.6	81	78	4.0	15.7	3.4
February	22.7	30.3	82	69	4.6	16.9	3.7
March	23.0	30.6	80	78	5.1	17.4	3.8
April	23.0	30.2	82	69	5.0	16.4	3.5
May	22.0	28.6	84	69	3.8	13.5	2.9
June	19.2	26.5	81	69	3.3	12.2	2.6
July	17.6	25.1	78	78	3.2	12.3	2.6
August	18.6	25.3	78	78	2.6	12.4	2.6
September	20.5	26.5	78	104	2.0	12.4	2.8
October	22.5	28.0	79	130	2.2	12.9	3.1
November	23.0	28.7	80	104	3.2	14.4	3.3
December	23.0	29.1	82	95	3.8	15.2	3.4
Year	21.5	28.2	80	85	3.6	14.3	3.1

Fig. ET_0 Computed by CROPWAT

Ten-Day or Monthly Time Step

Notwithstanding the non-linearity in the Penman-Monteith equation and some weather parameter methods, mean ten-day or monthly weather data can be used to compute the mean ten-day or monthly values for the reference evapotranspiration. The value of the reference evapotranspiration calculated with mean monthly weather data is indeed very similar to the average of the daily ET_0 values calculated with daily average weather data for that month.

The meteorological data consist of:

- Air temperature: ten-day or monthly average daily maximum (T_{max}) and average daily minimum temperature (T_{min}).
- Air humidity: ten-day or monthly average of the daily actual vapour pressure (e_a) derived from psychrometric, dewpoint or relative humidity data.
- Wind speed: ten-day or monthly average of daily wind speed data measured at 2 m height (u_2).
- Radiation: ten-day or monthly average of daily net radiation (R_n) computed from the mean ten-day or monthly measured shortwave radiation or from actual duration of daily sunshine hours (n). The extraterrestrial radiation (R_a) and daylight hours (N) for a specific day of the month can be computed.

When the soil is warming (spring) or cooling (autumn), the soil heat flux (G) for monthly periods may become significant relative to the mean monthly R_n. In these cases G cannot be ignored and its value should be determined from the mean monthly air temperatures of the previous and next month.

Daily Time Step

Calculation of ET_o with the Penman-Monteith equation on 24-hour time scales will generally provide accurate results. The required meteorological data consist of:

- *Air temperature:* Maximum (T_{max}) and minimum (T_{min}) daily air temperatures.
- *Air humidity:* Mean daily actual vapour pressure (e_a) derived from psychrometric, dewpoint temperature or relative humidity data.
- *Wind speed:* Daily average for 24 hours of wind speed measured at 2 m height (u_2).
- *Radiation:* Net radiation (R_n) measured or computed from solar and longwave radiation or from the actual duration of sunshine (n). The extraterrestrial radiation (R_a) and daylight hours (N) for a specific day of the month should be computed using Equations. As the magnitude of daily soil heat flux (G) beneath the reference grass surface is relatively small, it may be ignored for 24-hour time steps.

-	$0.408\,(R_n - G) =$	5.42	mm/day
Grass reference evapotranspiration			
-	$0.408\,(R_n - G)\,\Delta/[\Delta + \gamma\,(1 + 0.34\,u_2)] =$	2.81	mm/day
-	$900/(T+273)\,u_2\,(e_s - e_a)\,\gamma/[\Delta + \gamma\,(1+0.34\,u_2)] =$	1.07	mm/day
-	ET_o (Eq. 6) = 2.81 + 1.07 = 3.88 ≈	3.9	mm/day
The grass reference evapotranspiration is 3.9 mm/day.			

Hourly Time Step

In areas where substantial changes in wind speed, dewpoint or cloudiness occur during the day, calculation of the ET_o equation using hourly time steps is generally better than using 24-hour calculation time steps. Such weather changes can cause 24-hour means to misrepresent evaporative power of the environment during parts of the day and may introduce error into the calculations. However, under most conditions, application of the FAO Penman-Monteith equation with 24-hour data produces accurate results.

With the advent of electronic, automated weather stations, weather data are increasingly reported for hourly or shorter periods. Therefore, in situations where calculations are computerized, the FAO Penman-Monteith equation can be applied on an hourly basis with good results.

When applying the FAO Penman-Monteith equation on an hourly

or shorter time scale, the equation and some of the procedures for calculating meteorological data should be adjusted for the smaller time step. The FAO Penman-Monteith equation for hourly time steps is:

$$ET_o = \frac{0.408\Delta(R_n - G) + \gamma \frac{37}{T_{hr} + 273} u_2 (e^o(T_{hr}) - e_a)}{\Delta + \gamma(1 + 0.34u_2)}$$

where, ET_o reference evapotranspiration [mm hour^{-1}],
R_n net radiation at the grass surface [MJ m^{-2} hour^{-1}],
G soil heat flux density [MJ m^{-2} hour^{-1}],
T_{hr} mean hourly air temperature [°C],
Δ saturation slope vapour pressure curve at T_{hr} [kPa °C^{-1}],
γ psychrometric constant [kPa °C^{-1}],
$e°(T_{hr})$ saturation vapour pressure at air temperature T_{hr} [kPa],
e_a average hourly actual vapour pressure [kPa]
u_2 average hourly wind speed [m s^{-1}].

Given relative humidity measurements, the actual vapour pressure is determined as:

$$e_a = e^o(T_{hr}) \frac{RH_{hr}}{100}$$

where, e_a average hourly actual vapour pressure [kPa],
$e°(T_{hr})$ saturation vapour pressure at air temperature T_{hr} [kPa],
RH_{hr} average hourly relative humidity [%].

The net radiation is the difference between the net shortwave radiation (R_{ns}) and the net longwave radiation (R_{nl}) at the hourly time steps. Consequently:

- If R_{ns} and R_{nl} need to be calculated, the extraterrestrial radiation value (R_a) for the hourly period should be used.
- In the computation of R_{nl} by means of Equation, $\left(\sigma T_{max,K}{}^4 + \sigma T_{min,K}{}^4\right)/2$ is replaced by $\sigma_{Thr,k}{}^4$ and the Stefan-Boltzman constant becomes:
  σ = (4.903/24) 10^{-9} = 2.043 10^{-10} MJ m^{-2} hour^{-1}.

Since the ratio R_s/R_{so} is used to represent cloud cover, when calculating R_{nl} for hourly periods during the nighttime, the ratio R_s/R_{so} can be set equal to the R_s/R_{so} calculated for a time period occurring 2-3 hours before sunset, before the sun angle becomes small. This will generally serve as a good approximation of cloudiness occurring during the subsequent nighttime.

The hourly period that is 2 to 3 hours before sunset can be identified during computation of R_a as the period where w, calculated from Equation 31, is within the range (ω_s - 0.79) ≤ ω ≤ (ω_s - 0.52), where

ω_s is calculated using Equation 25. As a more approximate alternative, one can assume R_s/R_{so} = 0.4 to 0.6 during nighttime periods in humid and subhumid climates and R_s/R_{so} = 0.7 to 0.8 in arid and semiarid climates. A value of R_s/R_{so} = 0.3 presumes total cloud cover.

Soil heat flux is important for hourly calculations. Equations 45 and 46 can be used to derive G for the hourly periods.

The required meteorological data consist of:

- *Air temperature:* Mean hourly temperature (T_{hr}).
- *Air humidity:* Average hourly relative humidity (RH_{hr}).
- *Wind speed:* Average hourly wind speed data measured at 2 m height (u_2).
- *Radiation:* Total hourly solar (R_s) or net radiation (R_n).

Because of the need for standardization, the constants in Equation 53 presume a constant surface resistance (r_s) of 70 s/m during all periods. This constant resistance may cause some underprediction of hourly ET_o during some daytime periods when actual r_s may be somewhat lower. The constant resistance may cause some overprediction of hourly ET_o during evening periods when actual r_s may be somewhat higher.

However, when the calculations of hourly ET_o from Equation 53 are summed over 24 hour periods to produce an equivalent 24-hour ET_o, the hourly differences tend to compensate one another and the results are generally equivalent to calculations of ET_o made on a 24-hour time step. Precise estimates of ET_o for specific hourly periods may require the use of aerodynamic stability functions and functions for modifying the value for r_s based on levels of radiation, humidity and temperature.

Application of these functions are not normally required when hourly calculations are to be summed to 24-hour totals. Therefore, these functions are not described here.

Calculation Procedures with Missing Data

The meteorological data, required to estimate ET_o by means of the FAO Penman-Monteith equation, consist of air temperature, air humidity, wind speed and radiation. Where some of these data are missing or cannot be calculated, it is strongly recommended that the user estimate the missing climatic data with one of the procedures presented that the FAO Penman-Monteith method be used for the calculation of ET_o.

The use of an alternative ET_o calculation procedure, requiring only limited meteorological parameters, is less recommended.

Example illustrates the estimation of monthly ET_o with the FAO

Penman-Monteith for a data set containing only maximum and minimum air temperature. The procedures given to estimate missing humidity, radiation and wind speed data should be validated by comparing ET_o calculated with full and with limited data sets for weather stations in the region with complete data sets.

-	$0.408\ (R_n - G) =$	5.50	mm/day
Grass reference evapotranspiration:			
-	$0.408\ (R_n - G)\ \Delta / [\Delta + \gamma\ (1+0.34\ u_2)] =$	3.17	mm/day
-	$900/(T + 273)\ u_2\ (e_s - e_a)\ \gamma / [\Delta + \gamma\ (1+0.34\ u_2)] =$	1.39	mm/day
-	$ET_o = (3.17 + 1.39) =$	4.56	mm/day
The estimated grass reference evapotranspiration is 4.6 mm/day. If instead of 2 m/s, the wind speed is estimated as 1 or 3 m/s, ET_o would have been 7% lower (4.2 mm/day) or 6% higher (4.8 mm/day) respectively. In comparison, the Hargreaves equation (Equation 52) predicts ET_o = 5.0 mm/day			

PAN EVAPORATION METHOD

Pan Evaporation

The evaporation rate from pans filled with water is easily obtained. In the absence of rain, the amount of water evaporated during a period (mm/day) corresponds with the decrease in water depth in that period. Pans provide a measurement of the integrated effect of radiation, wind, temperature and humidity on the evaporation from an open water surface. Although the pan responds in a similar fashion to the same climatic factors affecting crop transpiration, several factors produce significant differences in loss of water from a water surface and from a cropped surface.

Reflection of solar radiation from water in the shallow pan might be different from the assumed 23% for the grass reference surface. Storage of heat within the pan can be appreciable and may cause significant evaporation during the night while most crops transpire only during the daytime. There are also differences in turbulence, temperature and humidity of the air immediately above the respective surfaces. Heat transfer through the sides of the pan occurs and affects the energy balance.

Notwithstanding the difference between pan-evaporation and the evapotranspiration of cropped surfaces, the use of pans to predict ET_o for periods of 10 days or longer may be warranted. The pan evaporation is related to the reference evapotranspiration by an empirically derived pan coefficient:

$$ET_o = K_p\ E_{pan}$$

where

ET_o reference evapotranspiration [mm/day],

K_p pan coefficient [-],

E_{pan} pan evaporation [mm/day].

Pan Coefficient (K_p)

Pan Types and Environment

Different types of pans exist. As the colour, size, and position of the pan have a significant influence on the measured results, the pan coefficients are pan specific.

In selecting the appropriate pan coefficient, not only the pan type, but also the ground cover in the station, its surroundings as well as the general wind and humidity conditions, should be checked.

The siting of the pan and the pan environment also influence the results. This is particularly so where the pan is placed in fallow rather than cropped fields.

Two cases are commonly considered: Case A where the pan is sited on a short green (grass) cover and surrounded by fallow soil; and Case B where the pan is sited on fallow soil and surrounded by a green crop.

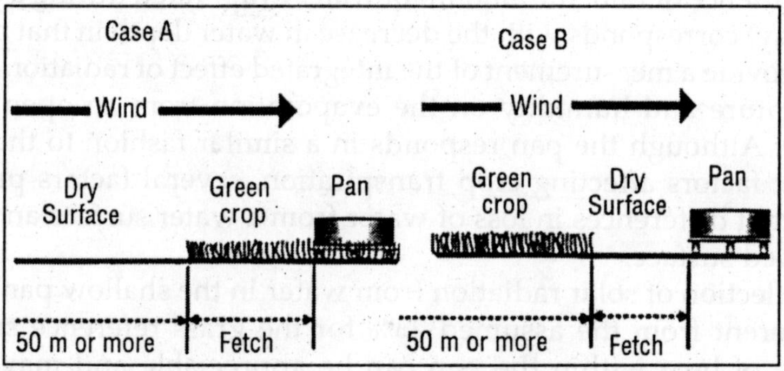

Fig. Two Cases of Evaporation Pan Siting and their Environment

Pan Coefficients

Depending on the type of pan and the size and state of the upwind buffer zone (fetch), pan coefficients will differ. The larger the upwind buffer zone, the more the air moving over the pan will be in equilibrium with the buffer zone.

At equilibrium with a large fetch, the air contains more water vapour and less heat in Case A than in Case B. Pan coefficients for the Class A pan and for the Colorado sunken pan for different ground cover, fetch and climatic conditions are presented in Tables.

Regression equations derived from the Tables are presented in Table. Where measured data from other types of sunken pans are available, such data should first be related to Colorado sunken pan data or to the FAO Penman-Monteith equation to develop K_p. Ratios between evaporation from sunken pans and from the Colorado sunken pan for different climatic conditions and pan environment are given in Table.

Where data are missing, wind speed can be estimated by taking a global value of $u_2 = 2$ m s^{-1} or as indicated in Table. RH_{mean} can be approximated from air temperature as $RH_{mean} = 50\ e°(T_{min})/e°(T_{max}) + 50$.

Adjustments

Under some conditions not accounted for in the Tables, the presented K_p coefficients may need some adjustment. This is the case in areas with no agricultural development, or where the pans are enclosed by tall crops. Not maintaining the standard colour of the pan or installing screens can affect the pan readings and will require some adjustment of the pan coefficient.

In areas with no agricultural development and extensive areas of bare soils (large fetch, Case B), as found under desert or semi-desert conditions, the listed values for K_p given for arid, windy areas may need to be reduced by up to 20%; for areas with moderate levels of wind, temperature and relative humidity, the listed values may need to be reduced by 5-10%; no or little reduction in K_p is needed in humid, cool conditions.

Where pans are placed in a small enclosure but surrounded by tall crops, for example 2.5 m high maize, the listed pan coefficients will need to be increased by up to 30% for dry windy climates whereas only a 5-10% increase is required for calm, humid conditions.

Painting the pans may affect the pan evaporation. The pan coefficients presented apply to galvanized pans annually painted with aluminium and to stainless steel pans.

Little difference in E_{pan} will occur where the inside and outside surfaces of the pan are painted white.

An increase in E_{pan} of up to 10% may occur when they are painted black. The material from which the pan is made may account for variations of only a few per cent.

The level at which the water is maintained in the pan is important; resulting errors may be up to 15% when water levels in the Class A pan fall 10 cm below the accepted standard of between 5 and 7.5 cm below the rim.

Screens mounted over pans will reduce E_{pan} by up to 10%. In an attempt to avoid pans being used by birds for drinking, pans filled to the rim with water can be placed near the Class A pan; birds may prefer to use the fully filled pan.

The evaporation pan should be placed in a large, secure, wire enclosure to prevent animals from entering and drinking. The turbidity of the water in the pan usually does not affect E_{pan} by more than 5%.

The overall variation in E_{pan} is not constant with time because of ageing, deterioration and repainting.

Table. Pan coefficients (K_p) for Class A pan for different pan siting and environment and different levels of mean relative humidity and wind speed

Class A pan	Case A: Pan placed in short green cropped area				Case B: Pan placed in dry fallow area			
RH mean (%) →		low < 40	me diu m 40 - 70	high > 70		low < 40	med ium 40 - 70	hig h > 70
Wind speed (m s^{-1})	Windw ard side distance of green crop (m)				Windward side distance of dry fallow (m)			
Light < 2	1	.55	.65	.75	1	.7	.8	.85
	10	.65	.75	.85	10	.6	.7	.8
	100	.7	.8	.85	100	.55	.65	.75
	1000	.75	.85	.85	1000	.5	.6	.7
Mode rate	1	.5	.6	.65	1	.65	.75.	.8
2-5	10	.6	.7	.75	10	.55	.65	.7
	100	.65	.75	.8	100	.5	.6	.65
	1000	.7	.8	.8	1000	.45	.55	.6.
Stron g	1	.45	.5	.6	1	.6	.65	.7
5-8	10	.55	.6	.65	10	.5	.55	.65
	100	.6	.65	.7	100	.45	.5	.6
	1000	.65	.7	.75	1000	.4	.45	.55
Very strong	1	.4	.45	.5	1	.5	.6	.65
> 8	10	.45	.55	.6	10	.45	.5	.55
	100	.5	.6	.65	100	.4	.45	.5
	1000	.55	.6	.65	1000	.35	.4	.45

Table. Pan coefficients (K_p) for Colorado sunken pan for different pan siting and environment and different levels of mean relative humidity and wind speed

Sunken Colorado	Case A: Pan placed in short green cropped area				Case B: Pan placed in dry fallow area (1)			
RH mean (%) →		low < 40	medium 40 - 70	high > 70		low < 40	medium 40 - 70	high > 70
Wind speed (m s^{-1})	Windward side distance of green crop (m)				Windward side distance of dry fallow (m)			
Light	1	.75	.75	.8	1	1.1	1.1	1.1
< 2	10	1.0	1.0	1.0	10	.85	.85	.85
	≥ 100	1.1	1.1	1.1	100	.75	.75	.8
					1000	.7	.7	.75
Moderate	1	.65	.7	.7	1	.95	.95	.95
2-5	10	.85	.85	.9	10	.75	.75	.75
	≥ 100	.95	.95	.95	100	.65	.65	.7
					1000	.6	.6	.65
Strong	1	.55	.6	.65	1	.8	.8	.8
5-8	10	.75	.75	.75	10	.65	.65	.65
	≥ 100	.8	.8	.8	100	.55	.6	.65
					1000	.5	.55	.6
Very strong	1	.5	.55	.6	1	.7	.75	.75
> 8	10	.65	.7	.7	10	.55	.6	.65
	≥ 100	.7	.75	.75	100	.5	.55	.6
					1000	.45	.5	.55

For extensive areas of bare-fallow soils and no agricultural development, reduce K_{pan} by 20% under hot, windy conditions; by 5-10% for moderate wind, temperature and humidity conditions.

Table. Pan coefficients (K_p): Regression Equations Derived from Tables

Class A pan with green fetch	$K_p = 0.108 - 0.0286\,u_2 + 0.0422\ln(FET) + 0.1434\ln(RH_{mean}) - 0.000631\,[\ln(FET)]^2\ln(RH_{mean})$
Class A pan with dry fetch	$K_p = 0.61 + 0.00341\,RH_{mean} - 0.000162\,u_2\,RH_{mean} - 0.00000959\,u_2\,FET + 0.00327\,u_2\ln(FET) - 0.00289\,u_2\ln(86.4\,u_2) - 0.0106\ln(86.4\,u_2)\ln(FET) + 0.00063\,[\ln(FET)]^2\ln(86.4\,u_2)$
Colorado sunken pan with green	$K_p = 0.87 + 0.119\ln(FET) - 0.0157[\ln(86.4\,u_2)]^2\ln(RH_{mean}) - 0.000053\ln(86.4u_2)\ln(FET)RH_{mean}$
Colorado sunken pan with dry fetch	$K_p = 1.145 - 0.080\,u_2 + 0.000903(u_2)^2\ln(RH_{mean}) - 0.0964\ln(FET) + 0.0031\,u_2\ln(FET) + 0.0015[\ln(FET)]^2\ln(RH_{mean})$
Coefficients and parameters	K_p pan coefficient [] u_2 average daily wind speed at 2 m height (m s^{-1}) RH_{mean} average daily relative humidity [%] = $(RH_{max} + RH_{min})/2$ FET fetch, or distance of the identified surface type (grass or short green agricultural crop for case A, dry crop or bare soil for case B upwind of the evaporation pan)
Range for variables	1 m $\leq$ FET $\leq$ 1000 m (these limits must be observed) $30\% \leq RH_{mean} \leq 84\%$ $1\ m\ s^{-1} \leq u_2 \leq 8\ m\ s^{-1}$

The above considerations and adjustments indicate that the use of Tables or the corresponding equations may not be sufficient to consider all local environmental factors influencing K_p and that local adjustment may be required. To do so, an appropriate calibration of E_{pan} against ET_o computed with the Penman-Monteith method is recommended.

It is recommended that the pan should be installed inside a short green cropped area with a size of a square of at least 15 by 15 m. The pan should not be installed in the centre but at a distance of at least 10 m from the green crop edge in the general upwind direction.

Where observations of wind speed and relative humidity, required for the computation of K_p, are not available at the site, estimates of the weather variables from a nearby station have to be utilized.

It is then recommended that these variables be averaged for the computation period and that E_{pan} be averaged for the same period.

Table. Ratios between the Eevaporation from Sunken Pans and a Colorado Sunken pan for different Climatic Conditions and Environments

		Ratio E_{pan} mentioned and E_{pan} Colorado			
Climate		Humid-temperate climate		Arid to semi-arid (dry season)	
Ground cover surrounding pan (50 m or more)		Short green cover	Dry fallow	Short green cover	Dry fallow
	Pan area (m^2)				
GGI 20 diameter 5 m, depth 2 m (former Soviet Union)	19.6	1.0	1.1	1.05	1.25
Sunken pan diameter 12 ft (3.66 m) depth 3.3 ft (Israel)	10.5				
BPI diameter 6 ft (1.83 m), depth 2 ft (0.61 m) (USA)	2.6				
Kenya pan diameter 4 ft (1.22 m) depth 14 in (0.356 m)	1.2				
Australian pan diameter 3 ft (0.91 m) depth 3 ft (0.91 m)	0.7		1.0		1.0

Symmons pan 6 ft² (0.56 m²) depth 2 ft (0.61 m)	0.6				
Aslyng pan 0.33 m², depth 1 m (Denmark)	0.3			1.0	
GGI 3000 diameter 0.618 cm, depth 60-80 cm (former Soviet Union)	0.3				
Sunken pan diameter 50 cm, depth 25 cm (Netherlands)	0.2	1.0	0.95	1.0	0.95

Determination of ET_o from Pan Evaporation using Tables

Given the daily evaporation data for the first week of July for a Class A pan installed in a green area surrounded by short irrigated field crops: 8.2, 7.5, 7.6, 6.8, 7.6, 8.9 and 8.5 mm/day. In that period the mean wind speed is 1.9 m/s and the daily mean relative humidity is 73%. Determine the 7-day average reference evapotranspiration.			
Pan is installed on a green surface: Case A			
Pan is surrounded by irrigated crops:	$fetch_{max}$ =	1000	m
Wind speed is light:	u <	2	m/s
Relative humidity is high:	RH_{mean} >	70	%
From Table 5 (for above conditions):	K_p =	0.85	-
	E_{pan} = (8.2 + 7.5 + 7.6 + 6.8 + 7.6 + 8.9 + 8.5)/7 =	7.9	mm/day
From Eq. 55:	ET_o = 0.85 (7.9) =	6.7	mm/day
The 7-day average of the crop reference evapotranspiration is 6.7 mm/day			

Description of Class A Pan

The Class A Evaporation pan is circular, 120.7 cm in diameter and

25 cm deep. It is made of galvanized iron (22 gauge) or Monel metal (0.8 mm). The pan is mounted on a wooden open frame platform which is 15 cm above ground level. The soil is built up to within 5 cm of the bottom of the pan. The pan must be level.

It is filled with water to 5 cm below the rim, and the water level should not be allowed to drop to more than 7.5 cm below the rim. The water should be regularly renewed, at least weekly, to eliminate extreme turbidity. The pan, if galvanized, is painted annually with aluminium paint. Screens over the pan are not a standard requirement and should preferably not be used. Pans should be protected by fences to keep animals from drinking.

The site should preferably be under grass, 20 by 20 m, open on all sides to permit free circulation of the air. It is preferable that stations be located in the centre or on the leeward side of large cropped fields.

Pan readings are taken daily in the early morning at the same time that precipitation is measured. Measurements are made in a stilling well that is situated in the pan near one edge. The stilling well is a metal cylinder of about 10 cm in diameter and some 20 cm deep with a small hole at the bottom.

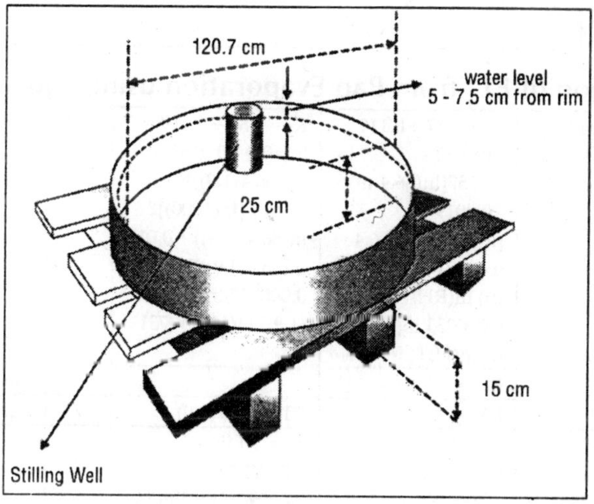

Description of Colorado Sunken Pan

The Colorado sunken pan is 92 cm (3 ft) square and 46 cm (18 in) deep, made of 3 mm thick iron, placed in the ground with the rim 5 cm (2 in) above the soil level. Also, the dimensions 1 m square and 0.5 m deep are frequently used. The pan is painted with black tar paint. The water level is maintained at or slightly below ground level, i.e., 5-7.5 cm below the rim.

Measurements are taken similarly to those for the Class A pan. Siting and environment requirements are also similar to those for the Class A pan. Sunken Colorado pans are sometimes preferred in crop water requirements studies, as these pans give a better direct estimation of the reference evapotranspiration than does the Class A pan. The disadvantage is that maintenance is more difficult and leaks are not visible.

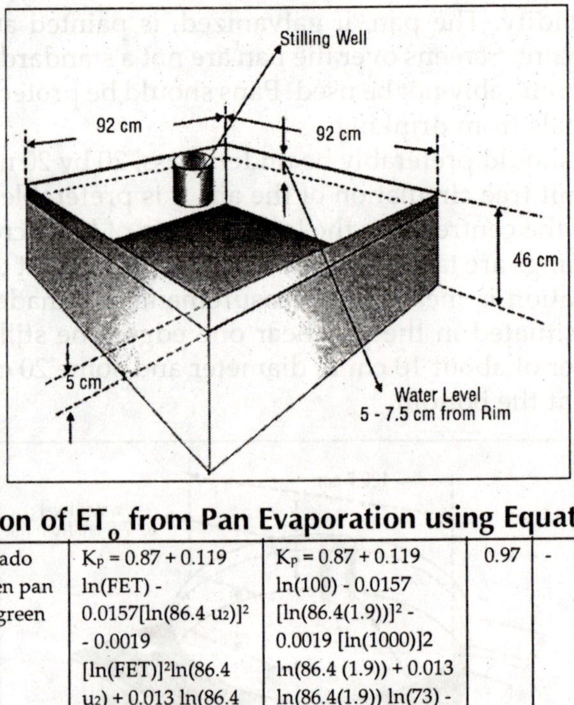

Determination of ET_o from Pan Evaporation using Equations

Colorado sunken pan with green fetch	K_p = 0.87 + 0.119 ln(FET) - 0.0157[ln(86.4 u_2)]^2 - 0.0019 [ln(FET)]^{2}ln(86.4 u_2) + 0.013 ln(86.4 u_2) ln(RH_{mean}) - 0.000053 ln(86.4 u_2)ln(FET) RH_{mean}	K_p = 0.87 + 0.119 ln(100) - 0.0157 [ln(86.4(1.9))]^2 - 0.0019 [ln(1000)]2 ln(86.4 (1.9)) + 0.013 ln(86.4(1.9)) ln(73) - 0.000053 ln(86.4 (1.9)) ln(1000) (73)	0.97	-
	E_{pan} =		7.9	mm/day
	ET_o =	ET_o = 0.97(7.9)	7.7	mm/day
Colorado sunken pan with dry fetch	K_p = 1.145 - 0.080 u_2 +0.000903(u_2)^{2}ln(RH_{mean}) - 0.0964 ln(FET) + 0.0031 u_2 ln(FET) + 0.0015 [ln(FET)]^2 ln(RH_{mean})	K_p = 1.145 - 0.080(1.9) + 0.000903(1.9)^{2}ln(73) - 0.0964 ln(1000) + 0.0031 (1.9) ln(1000) +0.0015 [ln(1000)]^{2}ln(73)	0.69	-
	E_{pan} =		7.9	mm/day
	ET_o =	ET_o = 0.69 (7.9)	5.4	mm/day
The 7-day average of the crop reference evapotranspiration for the four pan/fetch conditions is 6.6, 4.8, 7.7, and 5.4 mm/day				

Given the 7-day average evaporation measurement from Example 21, estimate the ET_o for the two types of pans and two types of fetch conditions represented by equations in Table 7. Assume that fetch is 1000 m in both fetch cases (green and dry).				
	Fetch =		1000	m
	u_2 =		1.9	m/s
	RH_{mean} =		73	%
Class A pan with green fetch	$K_p = 0.108 - 0.0286\ u_2 + 0.0422 \ln(FET) + 0.1434 \ln(RH_{mean}) - 0.000631\ [\ln(FET)]2\ \ln(RI\ I_{mean})$	$K_p = 0.108 - 0.0286\ (1.9) + 0.0422 \ln(1000) + 0.1434 \ln(73) - 0.000631\ [\ln(1000)]2 \ln(73)$	0.83	-
	E_{pan} =		7.9	mm/day
	ET_o =	$ET_o = 0.83\ (7.9)$	6.6	mm/day
Class A pan with dry fetch	$K_p = 0.61 + 0.00341\ RH_{mean} - 0.000162\ u_2\ RH_{mean} - 0.00000959\ u_2\ FET + 0.00327\ u_2 \ln(FET) - 0.00289\ u_2 \ln(86.4\ u_2) - 0.0106 \ln(86.4u_2)\ln(FET) + 0.00063\ [\ln(FET)]^2\ln(86.4\ u_2)$	$K_p = 0.61 + 0.00341\ (73) - 0.000162\ (1.9)\ (73) - 0.00000959\ (1.9)(1000) + 0.00327\ (1.9) \ln(1000) - 0.00289\ (1.9) \ln(86.4(1.9)) - 0.0106 \ln(86.4(1.9))\ln(1000) + 0.00063\ [\ln(1000)]^2 \ln(86.4\ (1.9))$	0.61	-
	E_{pan} =		7.9	mm/day
	ET_o =	$ET_o = 0.61\ (7.9)$	4.8	mm/day
Colorado sunken pan with green fetch	$K_p = 0.87 + 0.119 \ln(FET) - 0.0157[\ln(86.4\ u_2)]^2\ \ 0.0019\ [\ln(FET)]^2\ln(86.4\ u_2) + 0.013 \ln(86.4\ u_2) \ln(RH_{mean}) - 0.000053 \ln(86.4\ u_2)\ln(FET)\ RH_{mean}$	$K_p = 0.87 + 0.119 \ln(100) - 0.0157\ [\ln(86.4(1.9))]^2 - 0.0019\ [\ln(1000)]2 \ln(86.4\ (1.9)) + 0.013 \ln(86.4(1.9)) \ln(73) - 0.000053 \ln(86.4\ (1.9)) \ln(1000)\ (73)$	0.97	-
	E_{pan} =		7.9	mm/day
	ET_o =	$ET_o = 0.97(7.9)$	7.7	mm/day

Chapter 5

Single Crop Coefficient

This chapter deals with the calculation of crop evapotranspiration (ET_c) under standard conditions. No limitations are placed on crop growth or evapotranspiration from soil water and salinity stress, crop density, pests and diseases, weed infestation or low fertility. ET_c is determined by the crop coefficient approach whereby the effect of the various weather conditions are incorporated into ET_o and the crop characteristics into the K_c coefficient:

$$ET_c = K_c \, ET_o$$

The effect of both crop transpiration and soil evaporation are integrated into a single crop coefficient. The K_c coefficient incorporates crop characteristics and averaged effects of evaporation from the soil. For normal irrigation planning and management purposes, for the development of basic irrigation schedules, and for most hydrologic water balance studies, average crop coefficients are relevant and more convenient than the K_c computed on a daily time step using a separate crop and soil coefficient.

Only when values for K_c are needed on a daily basis for specific fields of crops and for specific years, must a separate transpiration and evaporation coefficient ($K_{cb} + K_e$) be considered.

The calculation procedure for crop evapotranspiration, ET_c, consists of:

- Identifying the crop growth stages, determining their lengths, and selecting the corresponding K_c coefficients;
- Adjusting the selected K_c coefficients for frequency of wetting or climatic conditions during the stage.
- Constructing the crop coefficient curve (allowing one to determine K_c values for any period during the growing period).
- Calculating ET_c as the product of ET_o and K_c.

Length of Growth Stages

FAO Irrigation and Drainage Paper provides general lengths for the four distinct growth stages and the total growing period for various types of climates and locations.

In some situations, the time of emergence of vegetation and the time of effective full cover can be predicted using cumulative degree-based regression equations or by more sophisticated plant growth models. These types of models should be verified or validated for the local area or for a specific crop variety using local observations.

Table. Lengths of Crop Development Stages* for Various Planting Periods and Climatic Regions (Days)

Crop	Init. (L_{ini})	Dev. (L_{dev})	Mid (L_{mid})	Late (L_{late})	Total	Plant Date	Region
a. Small Vegetables							
Cabbage	40	60	50	15	165	Sept	Calif. Desert, USA
Carrots	20	30	50/30	20	100	Oct/Jan	Arid climate
	30	40	60	20	150	Feb/Mar	Mediterranean
	30	50	90	30	200	Oct	Calif. Desert, USA
Cauliflower	35	50	40	15	140	Sept	Calif. Desert, USA
Celery	25	40	95	20	180	Oct	(Semi) Arid
	25	40	45	15	125	April	Mediterranean
	30	55	105	20	210	Jan	(Semi) Arid
Crucifers[1]	20	30	20	10	80	April	Mediterranean
	25	35	25	10	95	February	Mediterranean
	30	35	90	40	195	Oct/Nov	Mediterranean
Lettuce	20	30	15	10	75	April	Mediterranean
	30	40	25	10	105	Nov/Jan	Mediterranean
	25	35	30	10	100	Oct/Nov	Arid Region
	35	50	45	10	140	Feb	Mediterranean

	30	45	65	20	160	Dec/Jan	Arid Region
Water melons	20	30	30	30	110	April	Italy
	10	20	20	30	80	Mat/Aug	Near East (desert)
d. Roots and Tubers							
Beets, table	15	25	20	10	70	Apr/May	Mediterranean
	25	30	25	10	90	Feb/Mar	Mediterranean & Arid
Cassava: year 1	20	40	90	60	210	Rainy	Tropical regions
year 2	150	40	110	60	360	season	
Potato	25	30	30/45	30	115/130	Jan/Nov	(Semi) Arid Climate
	25	30	45	30	130	May	Continental Climate
	30	35	50	30	145	April	Europe
	45	30	70	20	165	Apr/May	Idaho, USA
	30	35	50	25	140	Dec	Calif. Desert, USA
Sweet potato	20	30	60	40	150	April	Mediterranean
	15	30	50	30	125	Rainy seas.	Tropical regions
Sugarbeet	30	45	90	15	180	March	Calif., USA
	25	30	90	10	155	June	Calif., USA
	25	65	100	65	255	Sept	Calif. Desert, USA
	50	40	50	40	180	April	Idaho, USA
	25	35	50	50	160	May	Mediterranean
	45	75	80	30	230	November	Mediterranean
	35	60	70	40	205	November	Arid Regions
e. Legumes *(Leguminosae)*							
Beans (green)	20	30	30	10	90	Feb/Mar	Calif., Mediterranean
	15	25	25	10	75	Aug/Sep	Calif., Egypt, Lebanon
Beans (dry)	20	30	40	20	110	May/June	Continental Climates
	15	25	35	20	95	June	Pakistan, Calif.
	25	25	30	20	100	June	Idaho, USA
Faba bean, broad bean	15	25	35	15	90	May	Europe
	20	30	35	15	100	Mar/Apr	Mediterranean
- dry	90	45	40	60	235	Nov	Europe
- green	90	45	40	0	175	Nov	Europe
Green gram, cowpeas	20	30	30	20	110	March	Mediterranean
Groundnut	25	35	45	25	130	Dry	West Africa
	35	35	35	35	140	season	High Latitudes
	35	45	35	25	140	May May/June	Mediterranean
Lentil	20	30	60	40	150	April	Europe
	25	35	70	40	170	Oct/Nov	Arid Region
Peas	15	25	35	15	90	May	Europe
	20	30	35	15	100	Mar/Apr	Mediterranean
	35	25	30	20	110	April	Idaho, USA
Soybeans	15	15	40	15	85	Dec	Tropics
	20	30/35	60	25	140	May	Central USA
	20	25	75	30	150	June	Japan
f. Perennial Vegetables (with winter dormancy and initially bare or mulched soil)							
Artichoke	40	40	250	30	360	Apr (1st yr)	California
	20	25	250	30	325	May (2nd yr)	(cut in May)
Asparagus	50	30	100	50	230	Feb	Warm Winter
	90	30	200	45	365	Feb	Mediterranean
g. Fibre Crops							
Cotton	30	50	60	55	195	Mar-May	Egypt; Pakistan; Calif.
	45	90	45	45	225	Mar	Calif. Desert, USA

	30	50	60	55	195	Sept	Yemen
	30	50	55	45	180	April	Texas
Flax	25	35	50	40	150	April	Europe
	30	40	100	50	220	October	Arizona
h. Oil Crops							
Castor beans	25	40	65	50	180	March	(Semi) Arid Climates
	20	40	50	25	135	Nov.	Indonesia
Safflower	20	35	45	25	125	April	California, USA
	25	35	55	30	145	Mar	High Latitudes
	35	55	60	40	190	Oct/Nov	Arid Region
Sesame	20	30	40	20	100	June	China
Sunflower	25	35	45	25	130	April/May	Medit.; California
i. Cereals							
Barley/Oats/Wheat	15	25	50	30	120	November	Central India
	20	25	60	30	135	March/Apr	35-45 °L
	15	30	65	40	150	July	East Africa
	40	30	40	20	130	Apr	
	40	60	60	40	200	Nov	
	20	50	60	30	160	Dec	Calif. Desert, USA
Winter Wheat	20[2]	60[2]	70	30	180	December	Calif., USA
	30	140	40	30	240	November	Mediterranean
	160	75	75	25	335	October	Idaho, USA
Grains (small)	20	30	60	40	150	April	Mediterranean
	25	35	65	40	165	Oct/Nov	Pakistan; Arid Reg.
Maize (grain)	30	50	60	40	180	April	East Africa (alt.)
	25	40	45	30	140	Dec/Jan	Arid Climate
	20	35	40	30	125	June	Nigeria (humid)
	20	35	40	30	125	October	India (dry, cool)
	30	40	50	30	150	April	Spain (spr, sum.); Calif.
	30	40	50	50	170	April	Idaho, USA
Maize (sweet)	20	20	30	10	80	March	Philippines
	20	25	25	10	80	May/June	Mediterranean
	20	30	50/30	10	90	Oct/Dec	Arid Climate
	30	30	30	103	110	April	Idaho, USA
	20	40	70	10	140	Jan	Calif. Desert, USA
Millet	15	25	40	25	105	June	Pakistan
	20	30	55	35	140	April	Central USA
Sorghum	20	35	40	30	130	May/June	USA, Pakis., Med.
	20	35	45	30	140	Mar/April	Arid Region
Rice	30	30	60	30	150	Dec; May	Tropics; Mediterranean
	30	30.	80	40	180	May	Tropics
j. Forages							
Alfalfa, total season	10	30	var.	var.	var.		last -4°C in spring until first -4°C in fall
Alfalfa 1st cutting cycle	10	20	20	10	60	Jan Apr (last - 4°C)	Calif., USA.
	10	30	25	10	75		Idaho, USA.
Alfalfa , other cutting cycles	5	10	10	5	30	Mar	Calif., USA.
	5	20	10	10	45	Jun	Idaho, USA.
Bermuda for seed	10	25	35	35	105	March	Calif. Desert, USA
Bermuda for hay (several cuttings)	10	15	75	35	135	---	Calif. Desert, USA
Grass Pasture	10	20	--	--	--		7 days before last -4°C in spring until 7

Sudan, 1st cutting cycle	25	25	15	10	75	Apr	Calif. Desert, USA
Sudan, other cutting cycles	3	15	12	7	37	June	Calif. Desert, USA
k. Sugar Cane							
Sugarcane, virgin	35	60	190	120	405		Low Latitudes
	50	70	220	140	480		Tropics
	75	105	330	210	720		Hawaii, USA
Sugarcane, ratoon	25	70	135	50	280		Low Latitudes
	30	50	180	60	320		Tropics
	35	105	210	70	420		Hawaii, USA
l. Tropical Fruits and Trees							
Banana, 1st yr	120	90	120	60	390	Mar	Mediterranean
Banana, 2nd yr	120	60	180	5	365	Feb	Mediterranean
Pineapple	60	120	600	10	790		Hawaii, USA
m. Grapes and Berries							
Grapes	20	40	120	60	240	April	Low Latitudes
	20	50	75	60	205	Mar	Calif., USA
	20	50	90	20	180	May	High Latitudes
	30	60	40	80	210	April	Mid Latitudes (wine)
Hops	25	40	80	10	155	April	Idaho, USA
n. Fruit Trees							
Citrus	60	90	120	95	365	Jan	Mediterranean
Deciduous Orchard	20	70	90	30	210	March	High Latitudes
	20	70	120	60	270	March	Low Latitudes
	30	50	130	30	240	March	Calif., USA
Olives	30	90	*60*	90	2705	March	Mediterranean
Pistachios	20	60	30	40	150	Feb	Mediterranean
Walnuts	20	10	130	30	190	April	Utah, USA
o. Wetlands - Temperate Climate							
Wetlands (Cattails, Bulrush)	10	30	80	20	140	May	Utah, USA; killing frost
	180	60	90	35	365	November	Florida, USA
Wetlands (short veg.)	180	60	90	35	365	November	frost-free climate

The lengths of the initial and development periods may be relatively short for deciduous trees and shrubs that can develop new leaves in the spring at relatively fast rates.

The rate at which vegetation cover develops and the time at which it attains effective full cover are affected by weather conditions in general and by mean daily air temperature in particular. Therefore, the length of time between planting and effective full cover will vary with climate, latitude, elevation and planting date. It will also vary with cultivar (crop variety). Generally, once the effective full cover for a plant canopy has been reached, the rate of further phenological development (flowering, seed development, ripening, and senescence) is more dependent on plant genotype and less dependent on weather. The variation in length of the growing period for one cultivar of rice for one region and for various planting dates.

The end of the mid-season and beginning of the late season is usually marked by senescence of leaves, often beginning with the lower leaves of plants. The length of the late season period may be relatively short (less than 10 days) for vegetation killed by frost (for example, maize at high elevations in latitudes > 40°N) or for agricultural crops that are harvested fresh (for example, Table beets and small vegetables).

High temperatures may accelerate the ripening and senescence of crops. Long duration of high air temperature (> 35°C) can cause some crops such as turf grass to go into dormancy. If severely high air temperatures are coupled with moisture stress, the dormancy of grass can be permanent for the remainder of the growing season. Moisture stress or other environmental stresses will usually accelerate the rate of crop maturation and can shorten the mid and late season growing periods. The values in Table are useful only as a general guide and for comparison purposes. The listed lengths of growth stages are average lengths for the regions and periods specified and are intended to serve only as examples. Local observations of the specific plant stage development should be used, wherever possible, to incorporate effects of plant variety, climate and cultural practices.

Local information can be obtained by interviewing farmers, ranchers, agricultural extension agents and local researchers, by conducting local surveys, or by remote sensing. When determining stage dates from local observations, the guidelines and visual descriptions may be helpful.

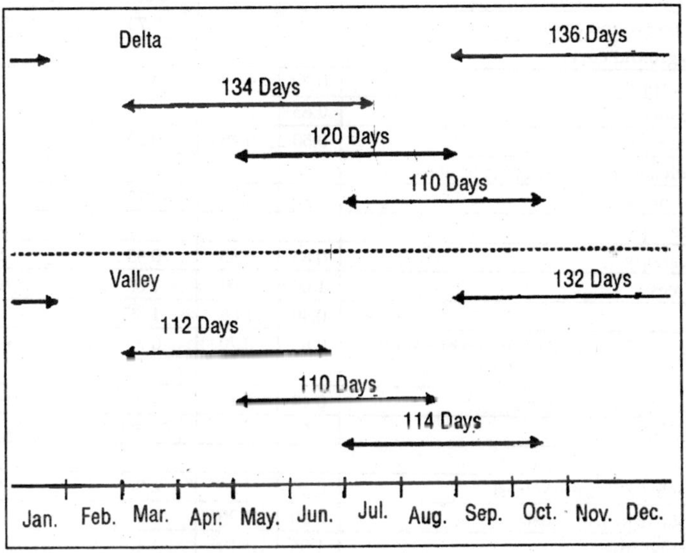

Fig. Variation in the Length of the Growing Period of Rice (cultivar: Jaya) Sown During Various Months of the Year at Different Locations along the Senegal River (Africa)

CROP COEFFICIENTS

Changes in vegetation and ground cover mean that the crop

coefficient K_c varies during the growing period. The trends in K_c during the growing period are represented in the crop coefficient curve. Only three values for K_c are required to describe and construct the crop coefficient curve: those during the initial stage ($K_{c\ ini}$), the mid-season stage ($K_{c\ mid}$) and at the end of the late season stage ($K_{c\ end}$).

Tabulated K_c Values

Typical values for $K_{c\ ini}$, Kc $_{mid}$ and $K_{c\ end}$ for various agricultural crops. The coefficients presented are organized by group type (i.e., small vegetables, legumes, cereals, etc.) to assist in locating the crop in the Table and to aid in comparing crops within the same group.

Table. Single (time-averaged) crop coefficients, K_c, and mean maximum plant heights for non stressed, well-managed crops in subhumid climates (RH_{min} » 45%, u_2 » 2 m/s) for use with the FAO Penman-Monteith ET_o.

Crop				
Apricots, Peaches, Stone Fruit [19, 20]				
- no ground cover, killing frost	0.45	0.90	0.65[18]	3
- no ground cover, no frosts	0.55	0.90	0.65[18]	3
- active ground cover, killing frost	0.50	1.15	0.90[18]	3
- active ground cover, no frosts	0.80	1.15	0.85[18]	3
Avocado, no ground cover	0.60	0.85	0.75	3
Citrus, no ground cover [21]				
- 70% canopy	0.70	0.65	0.70	4
- 50% canopy	0.65	0.60	0.65	3
- 20% canopy	0.50	0.45	0.55	2
Citrus, with active ground cover or weeds [22]				
- 70% canopy	0.75	0.70	0.75	4
- 50% canopy	0.80	0.80	0.80	3
- 20% canopy	0.85	0.85	0.85	2
Conifer Trees [23]	1.00	1.00	1.00	10
Kiwi	0.40	1.05	1.05	3
Olives (40 to 60% ground coverage by canopy) [24]	0.65	0.70	0.70	3-5
Pistachios, no ground cover	0.40	1.10	0.45	3-5
Walnut Orchard [19]	0.50	1.10	0.6518	4-5
o. Wetlands - temperate climate				
Cattails, Bulrushes, killing frost	0.30	1.20	0.30'	2
Cattails, Bulrushes, no frost	0.60	1.20	0.60	2
Short Veg., no frost	1.05	1.10	1.10	0.3
Reed Swamp, standing water	1.00	1.20	1.00	1-3
Reed Swamp, moist soil	0.90	1.20	0.70	1-3
p. Special				
Open Water, < 2 m depth or in subhumid climates or tropics		1.05	1.05	
Open Water, > 5 m depth, clear of turbidity, temperate climate		0.6525	1.2525	

Crop	$K_{c\,ini}$[1]	$K_{c\,mid}$	$K_{c\,end}$	Maximum Crop Height (h) (m)
a. Small Vegetables	0.7	1.05	0.95	
Broccoli		1.05	0.95	0.3
Brussel Sprouts		1.05	0.95	0.4
Cabbage		1.05	0.95	0.4
Carrots		1.05	0.95	0.3
Cauliflower		1.05	0.95	0.4
Celery		1.05	1.00	0.6
Garlic		1.00	0.70	0.3
Lettuce		1.00	0.95	0.3
Onions				
- dry		1.05	0.75	0.4
- green		1.00	1.00	0.3
- seed		1.05	0.80	0.5
Spinach		1.00	0.95	0.3
Radish		0.90	0.85	0.3
b. Vegetables - Solanum Family *(Solanaceae)*	0.6	1.15	0.80	
Egg Plant		1.05	0.90	0.8
Sweet Peppers (bell)		1.05[2]	0.90	0.7
Tomato		1.15[2]	0.70-0.90	0.6
c. Vegetables - Cucumber Family *(Cucurbitaceae)*	0.5	1.00	0.80	
Cantaloupe	0.5	0.85	0.60	0.3
Cucumber				
- Fresh Market	0.6	1.00[2]	0.75	0.3
- Machine harvest	0.5	1.00	0.90	0.3
Pumpkin, Winter Squash		1.00	0.80	0.4
Squash, Zucchini		0.95	0.75	0.3
Sweet Melons		1.05	0.75	0.4
Watermelon	0.4	1.00	0.75	0.4
d. Roots and Tubers	0.5	1.10	0.95	
Beets, table		1.05	0.95	0.4
Cassava				
- year 1	0.3	0.80[3]	0.30	1.0
- year 2	0.3	1.10	0.50	1.5
Parsnip	0.5	1.05	0.95	0.4
Potato		1.15	0.75[4]	0.6
Sweet Potato		1.15	0.65	0.4
Turnip (and Rutabaga)		1.10	0.95	0.6
Sugar Beet	0.35	1.20	0.70[5]	0.5
e. Legumes *(Leguminosae)*	0.4	1.15	0.55	
Beans, green	0.5	1.05[7]	0.90	0.4
Beans, dry and Pulses	0.4	1.15[2]	0.35	0.4
Chick pea		1.00	0.35	0.4
Fababean (broad bean)				
- Fresh	0.5	1.15[2]	1.10	0.8
- Dry/Seed	0.5	1.15[2]	0.30	0.8

Grabanzo	0.4	1.15	0.35	0.8
Green Gram and Cowpeas		1.05	0.60-0.35[6]	0.4
Groundnut (Peanut)		1.15	0.60	0.4
Lentil		1.10	0.30	0.5
Peas				
- Fresh	0.5	1.15[2]	1.10	0.5
- Dry/Seed		1.15	0.30	0.5
Soybeans		1.15	0.50	0.5-1.0
f. Perennial Vegetables (with winter dormancy and initially bare or mulched soil)	0.5	1.00	0.80	
Artichokes	0.5	1.00	0.95	0.7
Asparagus	0.5	0.95[7]	0.30	0.2-0.8
Mint	0.60	1.15	1.10	0.6-0.8
Strawberries	0.40	0.85	0.75	0.2
g. Fibre Crops	0.35			
Cotton		1.15-1.20	0.70-0.50	1.2-1.5
Flax		1.10	0.25	1.2
Sisal [8]		0.4-0.7	0.4-0.7	1.5
h. Oil Crops	0.35	1.15	0.35	
Castorbean (*Ricinus*)		1.15	0.55	0.3
Rapeseed, Canola		1.0-1.15[9]	0.35	0.6
Safflower		1.0-1.15[9]	0.25	0.8
Sesame		1.10	0.25	1.0
Sunflower		1.0-1.15[9]	0.35	2.0
i. Cereals	0.3	1.15	0.4	
Barley		1.15	0.25	1
Oats		1.15	0.25	1
Spring Wheat		1.15	0.25-0.4[10]	1
Winter Wheat				
- with frozen soils	0.4	1.15	0.25-0.4[10]	1
- with non-frozen soils	0.7	1.15	0.25-0.4[10]	
Maize, Field (grain) (*field corn*)		1.20	0.60-0.35[11]	2
Maize, Sweet (*sweet corn*)		1.15	1.05[12]	1.5
Millet		1.00	0.30	1.5
Sorghum				
- grain		1.00-1.10	0.55	1-2
- sweet		1.20	1.05	2-4
Rice	1.05	1.20	0.90-0.60	1
j. Forages				
Alfalfa Hay				
- averaged cutting effects	0.40	0.95[13]	0.90	0.7
- individual cutting periods	0.40[14]	1.20[14]	1.15[14]	0.7
- for seed	0.40	0.50	0.50	0.7

Bermuda hay				
- averaged cutting effects	0.55	1.00[13]	0.85	0.35
- Spring crop for seed	0.35	0.90	0.65	0.4
Clover hay, Berseem				
- averaged cutting effects	0.40	0.90[13]	0.85	0.6
- individual cutting periods	0.40[14]	1.15[14]	1.10[14]	0.6
Rye Grass hay				
- averaged cutting effects	0.95	1.05	1.00	0.3
Sudan Grass hay (annual)				
- averaged cutting effects	0.50	0.90[14]	0.85	1.2
- individual cutting periods	0.50[14]	1.15[14]	1.10[14]	1.2
Grazing Pasture				
- Rotated Grazing	0.40	0.85-1.05	0.85	0.15-0.30
- Extensive Grazing	0.30	0.75	0.75	0.10
Turf grass				
- cool season [15]	0.90	0.95	0.95	0.10
- warm season [15]	0.80	0.85	0.85	0.10
k. Sugar Cane	0.40	1.25	0.75	3
l. Tropical Fruits and Trees				
Banana				
- 1st year	0.50	1.10	1.00	3
- 2nd year	1.00	1.20	1.10	4
Cacao	1.00	1.05	1.05	3
Coffee				
- bare ground cover	0.90	0.95	0.95	2-3
- with weeds	1.05	1.10	1.10	2-3
Date Palms	0.90	0.95	0.95	8
Palm Trees	0.95	1.00	1.00	8
Pineapple [16]				
- bare soil	0.50	0.30	0.30	0.6-1.2
- with grass cover	0.50	0.50	0.50	0.6-1.2
Rubber Trees	0.95	1.00	1.00	10
Tea				
- non-shaded	0.95	1.00	1.00	1.5
- shaded [17]	1.10	1.15	1.15	2
m. Grapes and Berries				
Berries (bushes)	0.30	1.05	0.50	1.5
Grapes				
- Table or Raisin	0.30	0.85	0.45	2
- Wine	0.30	0.70	0.45	1.5-2
Hops	0.3	1.05	0.85	5
n. Fruit Trees				
Almonds, no ground cover	0.40	0.90	0.65[18]	5
Apples, Cherries, Pears [19]				
- no ground cover, killing frost	0.45	0.95	0.70[18]	4
- no ground cover, no frosts	0.60	0.95	0.75[18]	4
- active ground cover, killing frost	0.50	1.20	0.95[18]	4
- active ground cover, no frosts	0.80	1.20	0.85[18]	4

There is usually close similarity in the coefficients among the members of the same crop group, as the plant height, leaf area, ground coverage and water management are normally similar.

The coefficients in Table integrate the effects of both transpiration and evaporation over time. The effects of the integration over time represent an average wetting frequency for a 'standard' crop under typical growing conditions in an irrigated setting. The values for K_c during the initial and crop development stages are subject to the effects of large variations in wetting frequencies and therefore refinements to the value used for $K_{c\ ini}$ should always be made. For frequent wettings such as with high frequency sprinkler irrigation or rainfall, the values for $K_{c\ ini}$ may increase substantially.

The values for $K_{c\ mid}$ and $K_{c\ end}$ in Table represent those for a sub-humid climate with an average daytime minimum relative humidity (RH_{min}) of about 45% and with calm to moderate wind speeds averaging 2 m/s. For more humid or arid conditions, or for more or less windy conditions, the K_c coefficients for the mid-season and end of late season stage should be modified as described in this chapter.

The values for K_c in Table are values for non-stressed crops cultivated under excellent agronomic and water management conditions and achieving maximum crop yield (standard conditions). Where stand density, height or leaf area are less than that attained under such conditions, the value for $K_{c\ mid}$ and, for most crops, for $K_{c\ end}$ will need to be modified.

CROP COEFFICIENT FOR THE INITIAL STAGE ($K_{C\ INI}$)

Calculation Procedure

The values for $K_{c\ ini}$ are only approximations and should only be used for estimating ET_c during preliminary or planning studies. For several group types only one value for $K_{c\ ini}$ is listed and it is considered to be representative of the whole group for a typical irrigation water management. More accurate estimates of $K_{c\ ini}$ can be obtained by considering:

Time interval between wetting events

Evapotranspiration during the initial stage for annual crops is predominately in the form of evaporation. Therefore, accurate estimates for $K_{c\ ini}$ should consider the frequency with which the soil surface is wetted during the initial period. Where the soil is frequently wet from irrigation or rain, the evaporation from the soil surface can be considerable and $K_{c\ ini}$ will be large. On the other hand, where the soil surface is dry, evaporation is restricted and the $K_{c\ ini}$ will be small.

Evaporation Power of the Atmosphere

The value of $K_{c\ ini}$ is affected by the evaporating power of the atmosphere, i.e., ET_o. The higher the evaporation power of the atmosphere, the quicker the soil will dry between water applications and the smaller the time-averaged K_c will be for any particular period.

Magnitude of the wetting event

As the amount of water available in the topsoil for evaporation and hence the time for the soil surface to dry is a function of the magnitude of the wetting event, $K_{c\ ini}$ will be smaller for light wetting events than for large wettings.

Depending on the time interval between wetting events, the magnitude of the wetting event, and the evaporation power of the atmosphere, $K_{c\ ini}$ can vary between 0.1 and 1.15.

Time Interval between Wetting Events

In general, the mean time interval between wetting events is estimated by counting all rainfall and irrigation events occurring during the initial period that are greater than a few millimetres. Wetting events occurring on adjacent days can be counted as one event. The mean wetting interval is estimated by dividing the length of the initial period by the number of events.

Where only monthly rainfall values are available without any information on the number of rainy days, the number of events within the month can be estimated by dividing the monthly rainfall depth by the depth of a typical rain event.

The typical depth, if it exists, can vary widely from climate to climate, region to region and from season to season. Table presents some information on the range of rainfall depths. After deciding what rainfall is typical for the region and time of the year, the number of rainy days and the mean wetting interval can be estimated.

Table. Classification of Rainfall Depths

Rain Event	Depth
Very light (drizzle)	≤ 3 mm
Light (light showers)	5 mm
Medium (showers)	≥ 10 mm
Heavy (rainstorms)	≥ 40 mm

Where rainfall is insufficient, irrigation is needed to keep the crop well watered. Even where irrigation is not yet developed, the mean interval between the future irrigations should be estimated to obtain the required frequency of wetting necessary to keep the crop stress free.

The interval might be as small as a few days for small vegetables, but up to a week or longer for cereals depending on the climatic conditions. Where no estimate of the interval can be made, the user may refer to the values for $K_{c\ ini}$ of Table.

Example. Estimation of interval between wetting events

Estimate, from mean monthly rainfall data, the interval between rains during the rainy season for a station in a temperate climate (Paris, France: 50 mm/month), dry climate (Gafsa, Tunisia: 20 mm/month) and tropical climate (Calcutta, India: 300 mm/month).

Station	monthly rain (mm/month)	typical rainfall (mm)	number of rainy days	interval between rains
Paris	50	3	17	~ 2 days
Gafsa	20	5	4	weekly
Calcutta	300	20	15	~ 2 days

Determination of Kc ini

The crop coefficient for the initial growth stage can be derived which provide estimates for $K_{c\ ini}$ as a function of the average interval between wetting events, the evaporation power ET_o, and the importance of the wetting event.

Light wetting events (infiltration depths of 10 mm or less): rainfall and high frequency irrigation systems

When wetting during the initial period is only by precipitation, one will usually use to determine $K_{c\ ini}$. The graph can also be used when irrigation is by high frequency systems such as microirrigation and centre pivot and light applications of about 10 mm or less per wetting event are applied.

Graphical Determination of Kc ini

A silt loam soil receives irrigation every two days during the initial growth stage via a centre pivot irrigation system. The average depth applied by the centre pivot system is about 12 mm per event and the average ET_o during the initial stage is 4 mm/day. Estimate the crop evapotranspiration during that stage.

From Fig. 29 using the 2-day interval curve:	$K_{c\ ini}$ =	0.85	-
	$ET_c = K_c\ ET_o$ = 0.85 (4.0) =	3.4	mm/day

The average crop evapotranspiration during the initial growth stage is 3.4 mm/day

Heavy wetting events (infiltration depths of 40 mm or more): surface and sprinkler irrigation

For heavy wetting events when infiltration depths are greater than 40 mm, such as for when wetting is primarily by periodic irrigation such as by sprinkler or surface irrigation. Following a wetting event, the amount of water available in the topsoil for evaporation is considerable, and the time for the soil surface to dry might be significantly increased.

Consequently, the average K_c factor is larger than for light wetting events. As the time for the soil surface to dry is, apart from the evaporation power and the frequency of wetting, also determined by the water storage capacity of the topsoil, a distinction is made between soil types.

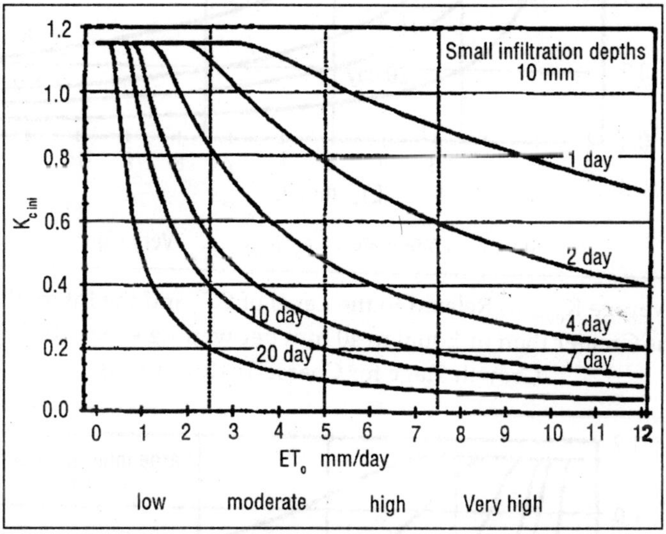

Fig. Average $K_{c\,ini}$ as related to the level of ET_o and the interval between irrigations and/or significant rain during the initial growth stage for all soil types when wetting events are light to medium (3-10 mm per event)

Coarse textured soils include sands and loamy sand textured soils. Medium textured soils include sandy loam, loam, silt loam and silt textured soils. Fine textured soils include silty clay loam, silty clay and clay textured soils.

Average wetting events (infiltration depths between 10 and 40 mm):

Where average infiltration depths are between 10 and 40 mm, the value for $K_{c\,ini}$ can be estimated:

$$K_{cini} = K_{cin} + \frac{(1-10)}{(40-10)}[K_c\,in - K_{cin}]$$

Where, $K_{c\,ini}$ value for $K_{c\,ini}$,

$K_{c\,ini}$ value for $K_{c\,ini}$,

I average infiltration depth [mm].

The values 10 and 40 are the average depths of infiltration (millimetres) upon which Figures are based.

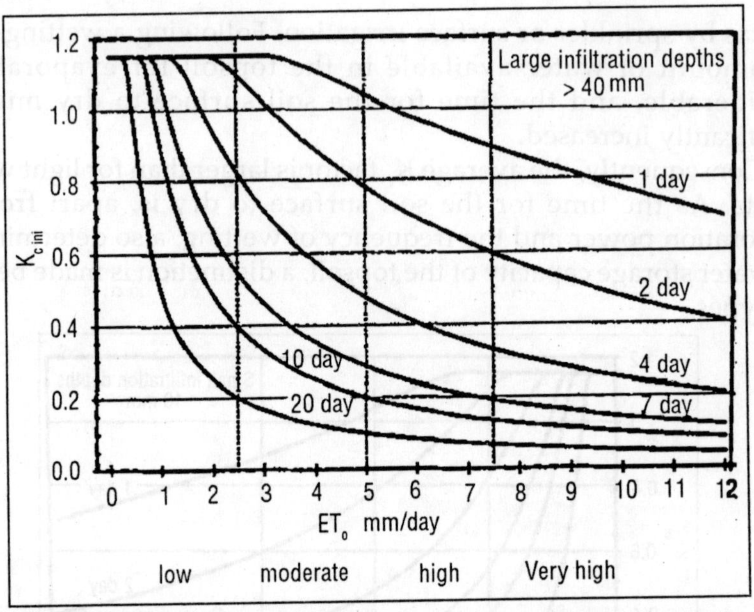

Fig. Average $K_{c\ ini}$ as Related to the Level of ET$_o$ and the Interval between Irrigations Greater than or Equal to 40 mm per wetting event, during the initial Growth Stage for Coarse Textured Soils

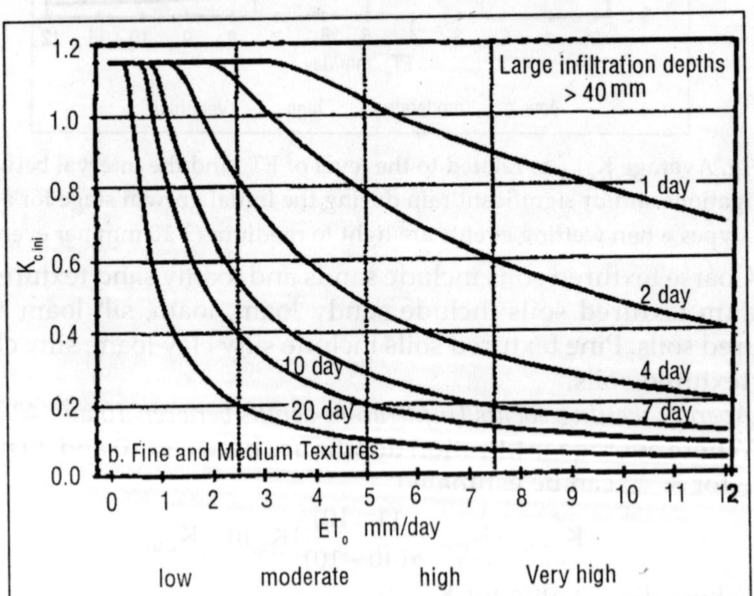

Adjustment for Partial Wetting by Irrigation

Many types of irrigation systems wet only a fraction of the soil surface. For example, for a trickle irrigation system, the fraction of the surface wetted, f_w, may be only 0.4.

For furrow irrigation systems, the fraction of the surface wetted may range from 0.3 to 0.8.

Example. Interpolation between light and heavy wetting events

Small vegetables cultivated in a dry area on a coarse textured soil receive 20 mm of water twice a week by means of a sprinkler irrigation system. The average ET_o during the initial stage is 5 mm/day. Estimate the crop evapotranspiration during that stage.			
For:	7/2=	3.5	day interval
	ET_o = and a coarse textured soil	5	mm/day
From Fig. 29:	$K_{c\,ini\,(Fig.\,29)}$?	0.55	-
From Fig. 30. a:	$K_{c\,ini\,(Fig.\,30a)}$?	0.7	-
For:	I =	20	mm
From Eq. 59:	$K_{c\,ini} = 0.55 + [(20 - 10)/(40 - 10)]\,(0.7 - 0.55) = 0.55 + 0.33(0.15) =$	0.60	
From Eq. 58:	$ET_c = 0.60\,(5) =$	3.0	mm/day
The average crop evapotranspiration during the initial growth stage for the small vegetables is 3.0 mm/day.			

Common values for the fraction of the soil surface wetted by irrigation or precipitation.

When only a fraction of the soil surface is wetted, the value for $K_{c\,ini}$ obtained should be multiplied by the fraction of the surface wetted to adjust for the partial wetting:

$$K_{c\,ini} = f_w\,K_{c\,ini}$$

where

f_w the fraction of surfaced wetted by irrigation or rain [0 - 1],

$K_{c\,ini}$ the value for $K_{c\,ini}$.

In addition, in selecting the average infiltrated depth, expressed in millimetres over the entire field surface, should be divided by f_w to represent the true infiltrated depth of water for the part of the surface that is wetted:

$$I_w = \frac{1}{f_w}$$

Where,

I_w irrigation depth for the part of the surface that is wetted [mm],

f_w fraction of surface wetted by irrigation,

I the irrigation depth for the field [mm].

When irrigation of part of the soil surface and precipitation over the entire soil surface both occur during the initial period, f_w should represent the average of f_w for each type of wetting, weighted according to the total infiltration depth received by each type.

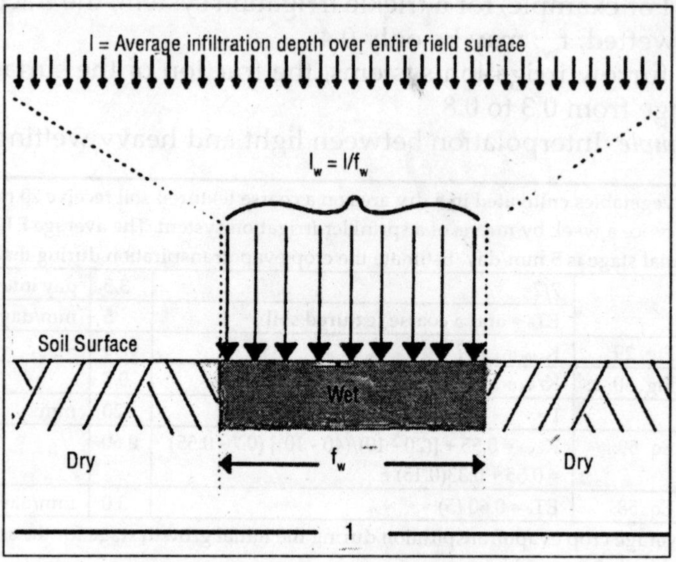

Fig. Partial Wetting by Irrigation

Kc ini for Trees and Shrubs

$K_{c\ ini}$ for trees and shrubs should reflect the ground condition prior to leaf emergence or initiation in case of deciduous trees or shrubs, and the ground condition during the dormancy or low active period for evergreen trees and shrubs. The $K_{c\ ini}$ depends upon the amount of grass or weed cover, frequency of soil wetting, tree density and mulch density.

For a deciduous orchard in frost-free climates, the $K_{c\ ini}$ can be as high as 0.8 or 0.9, where grass ground cover exists, and as low as 0.3 or 0.4 when the soil surface is kept bare and wetting is infrequent. The $K_{c\ ini}$ for an evergreen orchard (having no concerted leaf drop) with a dormant period has less variation from $K_{c\ mid}$, as exemplified for citrus. For 50% canopy or less, the $K_{c\ ini}$ also reflects ground cover conditions (bare soil, mulch or active grass or weed cover).

Kc ini for Paddy Rice

For rice growing in paddy fields with a water depth of 0.10-0.20 m, the ET_c during the initial stage mainly consists of evaporation from the standing water. The $K_{c\ ini}$ in table is a sub-humid climate with calm to moderate wind speeds. The $K_{c\ ini}$ should be adjusted for the local climate as indicated in table.

Table. $K_{c\ ini}$ for Rice for Various Climatic Conditions

Humidity	Wind speed		
	light	moderate	strong
arid - semi-arid	1.10	1.15	1.20
sub-humid - humid	1.05	1.10	1.15
very humid	1.00	1.05	1.10

Crop Coefficient for the Mid-season Stage ($K_{c\ mid}$)

Illustration of the Climatic Effect

Typical values for the crop coefficient for the mid-season growth stage, $K_{c\ mid}$, are listed in Table for various agricultural crops.

The effect of the difference in aerodynamic properties between the grass reference surface and agricultural crops is not only crop specific but also varies with the climatic conditions and crop height. More arid climates and conditions of greater wind speed will have higher values for $K_{c\ mid}$.

More humid climates and conditions of lower wind speed will have lower values for $K_{c\ mid}$.

The relative impact of climate on $K_{c\ mid}$ where the adjustments to the values from Table are shown for various types of climates, mean daily wind speeds and various crop heights. As an example, expected variations for $K_{c\ mid}$ for tomatoes in response to regional climatic conditions are presented.

Determination of Kc Mid

For specific adjustment in climates where RH_{min} differs from 45% or where u_2 is larger or smaller than 2.0 m/s, the $K_{c\ mid}$ values from Table are adjusted as:

$$K_{cmid} = K_{cmid}(Tab) + [0.04(u_2 - 2) - 0.004(RH_{min} - 45)]\left(\frac{h}{3}\right)^{0.3}$$

where, $K_{c\ mid\ (Tab)}$ value for $K_{c\ mid}$,

u_2 mean value for daily wind speed at 2 m height over grass during the mid-season growth stage [m s^{-1}], for 1 m $s^{-1} < u_2 \le 6$ m s^{-1},

RH_{min} mean value for daily minimum relative humidity during the mid-season growth stage [%], for $20\% \le RH_{min} \le 80\%$,

h mean plant height during the mid-season stage [m] for 0.1 m < h < 10 m.

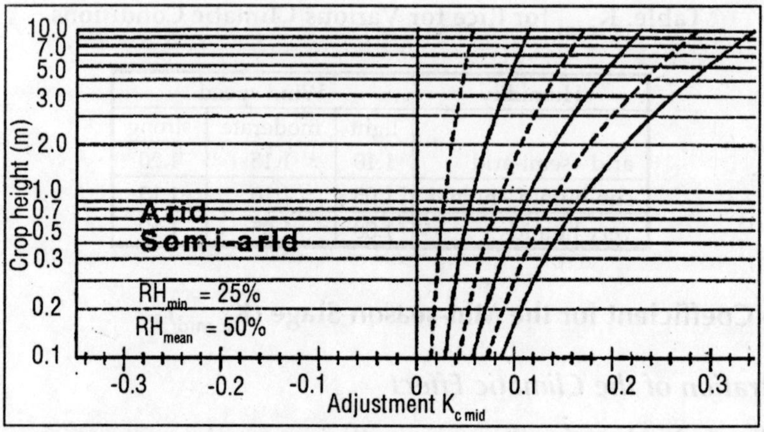

Fig. Adjustment (additive) to the $K_{c\ mid}$ values from Table for Different Crop Heights and Mean Daily Wind Speeds (u_2) for Different Humidity Conditions

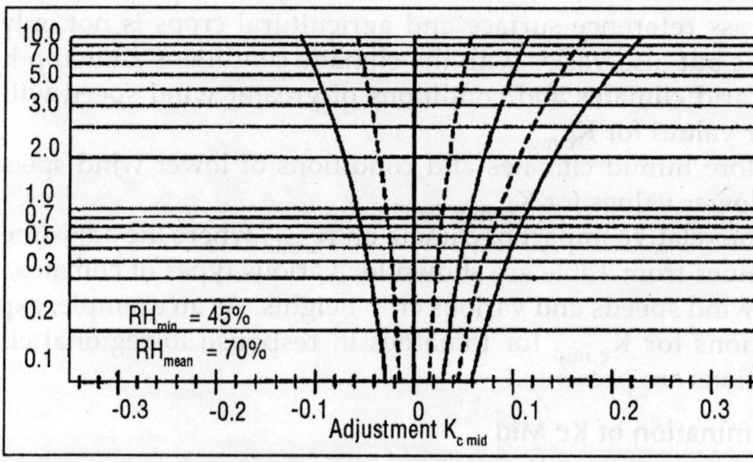

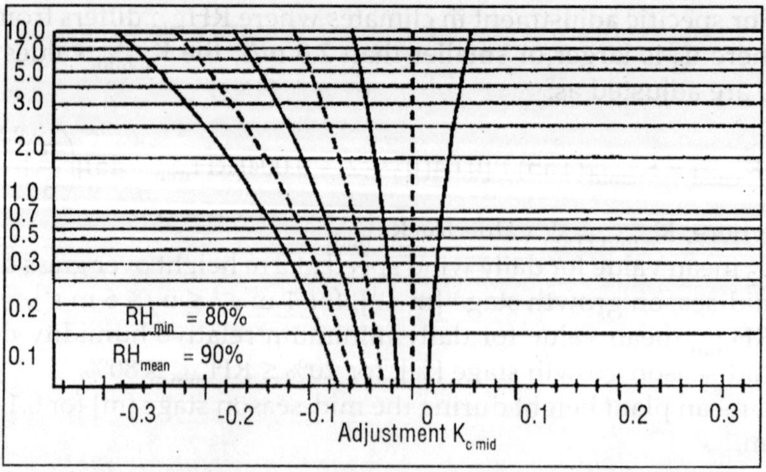

The $K_{c\ mid}$ values determined with equations are average adjustments for the midseason and late season periods. The values for parameters u_2 and RH_{min} should be accordingly taken as averages for these periods.

The limits expressed for parameters u_2, RH_{min} and h should be observed.

Where the user does not have access to a calculator with an exponential function, the solution of the $(h/3)^{0.3}$ expression can be approximated as $[(h/3)^{0.5}]^{0.5}$ where the square root key is used.

RH_{min} is used rather than RH_{mean} because it is easier to approximate RH_{min} from T_{max} where relative humidity data are unavailable.

Moreover, under the common condition where T_{min} approaches T_{dew} (i.e., $RH_{max} \approx 100\%$), the vapour pressure deficit $(e_s - e_a)$, with e_s from Equation and e_a from Equation, becomes $[(100 - RH_{min})/200]$ $e°(T_{max})$, where $e°(T_{max})$ is saturation vapour pressure at maximum daily air temperature. This indicates that RH_{min} better reflects the impact of vapour pressure deficit on K_c than does RH_{mean}.

RH_{min} is calculated on a daily or average monthly basis as:

$$RH_{min} = \frac{e^o(T_{dew})}{e^o(T_{max})} 100$$

where T_{dew} is mean dewpoint temperature and T_{max} is mean daily maximum air temperature during the mid-season growth stage. Where dewpoint temperature or other hygrométrie data are not available or are of questionable quality, RH_{min} can be estimated by substituting mean daily minimum air temperature, T_{min}, for T_{dew}[1]. Then:

$$RH_{min} = \frac{e^o(T_{min})}{e^o(T_{max})} 100$$

In the case of arid and semi-arid climates, T_{min} in equation should be adjusted by subtracting 2°C from the average value of T_{min} to better approximate T_{dew}.

The values for u_2 and RH_{min} need only be approximate for the mid-season growth stage. This is because Equation is not strongly sensitive to these values, changing 0.04 per 1 m/s change in u_2 and per 10% change in RH_{min} for a 3 m tall crop.

Measurements, calculation, and estimation of missing wind and humidity.

Wind speed measured at other than 2 m height should be adjusted to reflect values for wind speed at 2 m over grass. Where no data on u_2 or RH_{min} are available, the general classification for wind speed and humidity can be used.

Table. Empirical Estimates of Monthly Wwind Speed Data

Description	Mean Monthly Wind Speed at 2 m
light wind	...$\leq$ 1.0 m/s
light to moderate wind	2.0 m/s
moderate to strong wind	4.0 m/s
strong wind	... $\geq$ 5.0 m/s
general global conditions	2 m/s

Table. Typical values for RH_{min} compared with RH_{mean} for general climatic classifications

Climatic Classification	RH_{min} (%)	RH_{mean} (%)
Arid	20	45
Semi-arid	30	55
Sub-humid	45	70
Humid	70	85
Very humid	80	90

Equation is valid for mean plant heights up to 10 m. For plant heights smaller than 0.1 m, vegetation will behave aerodynamically similar to grass should not be applied. However, the mean plant height will greatly vary with crop variety and with cultural practices. Therefore, wherever possible, h should be obtained from general field observations.

However, the presence of the 0.3 exponent in Equation makes these equations relatively insensitive to small errors in the value used for h. Generally, a single value for h is used to represent me mid-season period.

Adjustment for Frequency of Wetting

$K_{c\ mid}$ is less affected by wetting frequency than is $K_{c\ ini}$, as vegetation during this stage is generally near full ground cover so that the effects of surface evaporation on K_c are smaller.

For frequent irrigation of crops (more frequently than every 3 days) and where the $K_{c\ mid}$ of Table is less than 1.0, the value can be replaced by approximately 1.1 1.3 to account for the combined effects of continuously wet soil, evaporation due to interception (sprinkler irrigation) and roughness of the vegetation, especially where the irrigation system moistens an important fraction of the soil surface (f_w > 0.3).

Determination of $K_{c\ mid}$

Calculate $K_{c\ mid}$ for maize crops near Taipei, Taiwan and near Mocha, Yemen. The average mean daily wind speed (u_2) during the mid-season stage at Taipei is about 1.3 m/s and the minimum relative humidity (RH_{min}) during this stage averages 75%. The average u_2 during the mid-season near Mocha is 4.6 m/s and the RH_{min} during this stage averages 44%.

From Table, the value for $K_{c\,mid}$ is 1.20 for maize. The value for h from Table is 2 m. Using Equation

For Taipei (humid climate):

$$K_{cmid} = 1.20 + [0.04(1.3-2) - 0.004(75-45)]\left(\frac{2}{3}\right)^{0.3} = 1.07$$

For Mocha (arid climate):

$$K_{cmid} = 1.20 + [0.04(4.6-2) - 0.004(44-45)]\left(\frac{2}{3}\right)^{0.3} = 1.30$$

The average crop coefficient predicted during the mid-season stage is 1.07 for Taipei and 1.30 for Mocha.

Crop Coefficient for the end of the Late Season Stage ($K_{c\,end}$)

Typical values for the crop coefficient at the end of the late season growth stage, $K_{c\,end}$, and listed in Table for various agricultural crops. The values given for $K_{c\,end}$ reflect crop and water management practices particular to those crops.

If the crop is irrigated frequently until harvested fresh, the topsoil remains wet and the $K_{c\,end}$ value will be relatively high. On the other hand, crops that are allowed to senesce and dry out in the field before harvest receive less frequent irrigation or no irrigation at all during the late season stage. Consequently, both the soil surface and vegetation are dry and the value for $K_{c\,end}$ will be relatively small.

Where the local water management and harvest timing practices are known to deviate from the typical values presented, then the user should make some adjustments to the values for $K_{c\,end}$. Some guidance on adjustment of K_c values for wetting frequency is provided. For premature harvest, the user can construct a K_c curve using the $K_{c\,end}$ value provided and a late season length typical of a normal harvest date; but can then terminate the application of the constructed curve early, corresponding to the time of the early harvest.

The $K_{c\,end}$ values are typical values expected for average $K_{c\,end}$ under the standard climatic conditions. More arid climates and conditions of greater wind speed will have higher values for $K_{c\,end}$. More humid climates and conditions of lower wind speed will have lower values for $K_{c\,end}$. For specific adjustment in climates where RH_{min} differs from 45% or where u_2 is larger or smaller than 2.0 m/s, Equation can be used:

$$Kcend = Kcend(Tab) + [0.04(u_2 - 2) - 0.004(RH_{min} - 45)]\left(\frac{h}{3}\right)^{0.3}$$

where, $K_{c\ end\ (Tab)}$ value for $K_{c\ end}$ taken from Table,

u_2 mean value for daily wind speed at 2 m height over grass during the late season growth stage [m s^{-1}], for 1 m $s^{-1} \le u_2 \le 6$ m s^{-1},

RH_{min} mean value for daily minimum relative humidity during the late season stage [%], for $20\% \le RH_{min} \le 80\%$,

h mean plant height during the late season stage [m], for 0.1 m $\le$ h $\le$ 10 m.

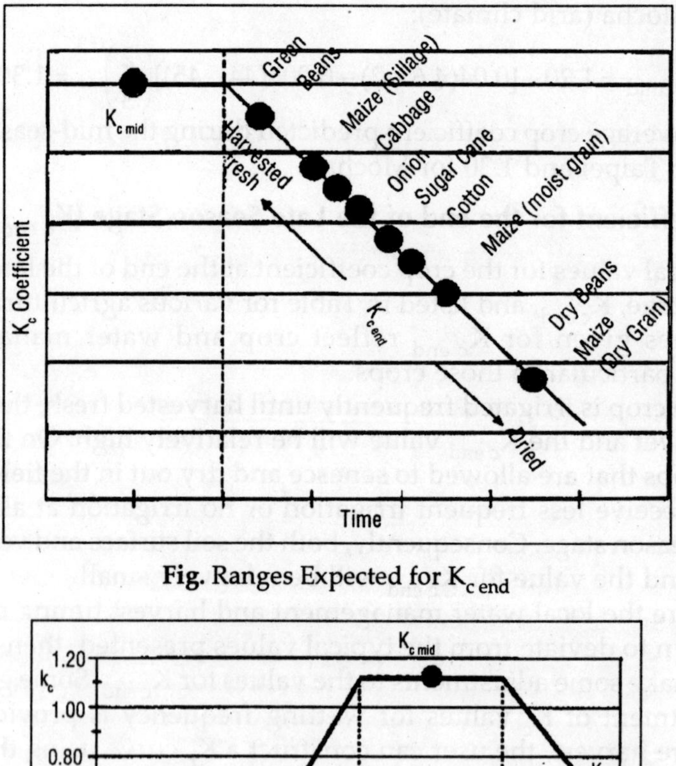

Fig. Ranges Expected for $K_{c\ end}$

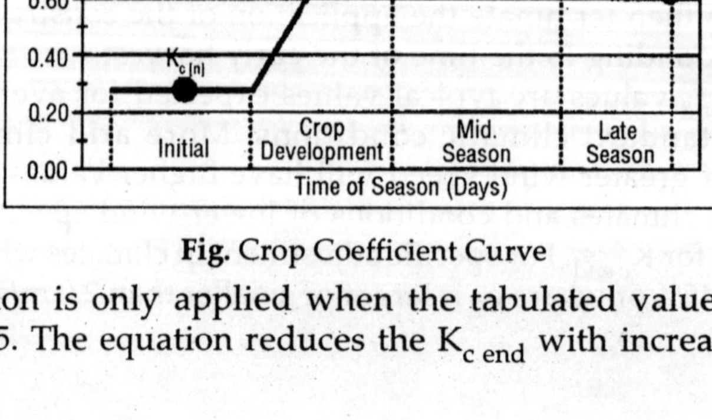

Fig. Crop Coefficient Curve

Equation is only applied when the tabulated values for $K_{c\ end}$ exceed 0.45. The equation reduces the $K_{c\ end}$ with increasing RH_{min}.

This reduction in $K_{c\ end}$ is characteristic of crops that are harvested 'green' or before becoming completely dead and dry (i.e., $K_{c\ end} \geq 0.45$).

No adjustment is made when $K_{c\ end\ (Table)} < 0.45$ (i.e., $K_{c\ end} = K_{c\ end\ (Tab)}$). When crops are allowed to senesce and dry in the field (as evidenced by $K_{c\ end} < 0.45$), u_2 and RH_{min} have less effect on $K_{c\ end}$ and no adjustment is necessary. In fact, $K_{c\ end}$ may decrease with decreasing RH_{min} for crops that are ripe and dry at the time of harvest, as lower relative humidity accelerates the drying process.

Construction of the K_c Curve

Annual Crops

Only three point values for K_c are required to describe and to construct the K_c curve. The curve is constructed using the following three steps:

- Divide the growing period into four general growth stages that describe crop phenology or development (initial, crop development, mid-season, and late season stage), determine the lengths of the growth stages, and identify the three K_c values that correspond to $K_{c\ ini}$, $K_{c\ mid}$ and $K_{c\ end}$.
- Adjust the K_c values to the frequency of wetting and/or climatic conditions of the growth stages as outlined in the previous section.
- Construct a curve by connecting straight line segments through each of the four growth stages. Horizontal lines are drawn through $K_{c\ ini}$ in the initial stage and through $K_{c\ mid}$ in the mid-season stage. Diagonal lines are drawn from $K_{c\ ini}$ to $K_{c\ mid}$ within the course of the crop development stage and from $K_{c\ mid}$ to $K_{c\ end}$ within the course of the late season stage.

K_c Curves for Forage Crops

Many crops grown for forage or hay are harvested several times during the growing season. Each harvest essentially terminates a 'sub' growing season and associated K_c curve and initiates a new 'sub' growing season and associated K_c curve.

The resulting K_c curve for the entire growing season is the aggregation of a series of K_c curves associated with each sub-cycle. figure presents a K_c curve for the entire growing season constructed for alfalfa grown for hay in southern Idaho.

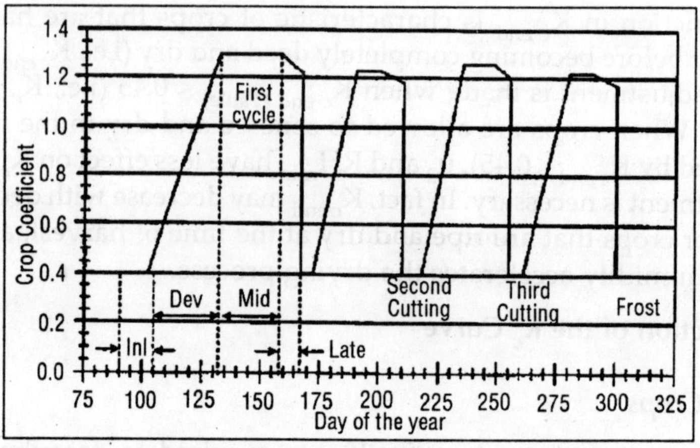

Fig. Constructed Curve for K_c for Alfalfa hay

In the southern Idaho climate, greenu› (leaf initiation) begins in the spring on about day 90 of the year. Tl.e crop is usually harvested (cut) for hay three or four times during t.ie growing season.

Therefore, four K_c sub-cycles or cutt.ng cycles: sub-cycle 1 follows greenup in the spring and the three additional K_c sub-cycles follow cuttings. Cuttings create a ground surface with less than 10% vegetation cover. Cutting cycle 1 is longer in duration than cycles 2, 3 and 4 due to lower air and soil temperatures during this period that reduce crop growth rates.

The lengths for cutting cycle 1 were taken from the first entry for alfalfa (" 1st cutting cycle") in Table for Idaho, the United States (10/30/25/10). The lengths for cutting cycles 2, 3 and 4 were taken from the entry for alfalfa in Table for "individual cutting periods" for Idaho, the United States (5/20/10/10). These lengths were based on observations. In the southern Idaho climate, frosts terminate the growing season sometime in the fall, usually around day 280-290 of the year (early to mid-October).

The magnitudes of the K_c values during the mid-season periods of each cutting cycle vary from cycle to cycle due to the effects of adjusting the values for $K_{c\ mid}$ and $K_{c\ end}$ for each cutting cycle period using Equations. In applying these two adjustment equations, the u_2 and RH_{min} values were averages for the mid-season and late season stages within each cutting cycle.

Kc mid when Effects of Individual Cutting Periods are Averaged

Under some conditions, the user may wish to average the effects of cuttings for a forage crop over the course of the growing season. When cutting effects are averaged, then only a single value for $K_{c\ mid}$

and a only single K_c curve need to be employed for the whole growing season. When this is the case, a "normal" K_c curve is constructed, where only one midseason period is shown for the forage crop. The $K_{c\ mid}$ for this total midseason period must average the effects of occasional cuttings or harvesting.

The value that is used for $K_{c\ mid}$ is therefore an average of the K_c curve for the time period starting at the first attainment of full cover and ending at the beginning of the final late season period near dormancy or frost. The value used for $K_{c\ mid}$ under these averaged conditions may be only about 80% of the K_c value that represents full ground cover. These averaged, full-season $K_{c\ mid}$ values are listed in Table. For example, for alfalfa hay, the averaged, full-season $K_{c\ mid}$ is 1.05, whereas, the $K_{c\ mid}$ for an individual cutting period is 1.20.

Fruit Trees

Values for the crop coefficient during the mid-season and end of late season stages. As mentioned before, the K_c values listed are typical values for standard climatic conditions and need to be adjusted by using Equations where RH_{min} or u_2 differ. As the mid and late season stages of deciduous trees are quite long, the specific adjustment of K_c to RH_{min} and u_2 should take into account the varying climatic conditions throughout the season.

Therefore, several adjustments of K_c are often required if the mid and late seasons cover several climatic seasons, e.g., spring, summer and autumn or wet and dry seasons. The $K_{c\ ini}$ and $K_{c\ end}$ for evergreen non dormant trees and shrubs are often not different, where climatic conditions do not vary much, as happens in tropical climates. Under these conditions, seasonal adjustments for climate may therefore not be required since variations in ET_c depend mostly on variations in ET_o.

Calculating ET_c

From the crop coefficient curve the K_c value for any period during the growing period can be graphically or numerically determined. Once the K_c values have been derived, the crop evapotranspiration, ET_c, can be calculated by multiplying the K_c values by the corresponding ET_o values.

Graphical Determination of K_c

Weekly, ten-day or monthly values for K_c are necessary when ET_c calculations are made on weekly, ten-day or monthly time steps. A general procedure is to construct the K_c curve, overlay the curve with the lengths of the weeks, decade or months, and to derive graphically from the curve the K_c value for the period under consideration. Assuming that all decades have a duration of 10 days facilitates the derivation of K_c and introduces little error into the calculation of ET_c.

This curve has been overlaid with the lengths of the decades. K_c values of 0.15, 1.19 and 0.35 and the actual lengths for growth stages equal to 25, 25, 30 and 20 days were used. The crop was planted at the beginning of the last decade of May and was harvested 100 days later at the end of August.

For all decades the K_c values can be derived directly from the curve. The value at the middle of the decade is considered to be the average K_c of that 10 day period. Only the second decade of June, where the K_c value changes abruptly, requires some calculation.

Case study of a dry bean crop at Kimberly, Idaho, the United States (single crop coefficient)

An example application for using the K_c procedure under average soil wetness conditions is presented for a dry bean crop planted on 23 May 1974 at Kimberly, Idaho, the United States (latitude = 42.4°N).

The initial, development, mid-season and late season stage lengths are taken from Table for a continental climate as 20, 30, 40 and 20 days (the stage lengths listed for southern Idaho were not used in this example in order to demonstrate the only approximate accuracy of values provided in Table when values for the specific location are not available). Initial values for $K_{c\,ini}$, $K_{c\,mid}$ and $K_{c\,end}$ are selected from Table

The mean RH_{min} and u_2 during both the mid-season and late season growth stages were 30% and 2.2 m/s. The maximum height suggested in Table for dry beans is 0.4 m. Therefore, $K_{c\,mid}$ is adjusted using Equation as:

$$K_{cmid} = 1.15 + [0.04(2.2-2) - 0.004(30-45)]\left(\frac{0.4}{3}\right)^{0.3} = 1.19$$

As $K_{c\,end} = 0.35$ is less than 0.45, no adjustment is made to $K_{c\,end}$. The value for $K_{c\,mid}$ is not significantly different from that in Table as ≈2 m/s, RH_{min} is just 15% lower than the 45% represented in Table, and the height of the beans is relatively short.

The initial K_c curve for dry beans in Idaho can be drawn, for initial, planning purposes, as shown in the graph (dotted line), where $K_{c\,ini}$, $K_{c\,mid}$ and $K_{c\,end}$ are 0.4, 1.19 and 0.35 and the four lengths of growth stages are 20, 30, 40 and 20 days. Note that the $K_{c\,ini} = 0.4$ taken from Table serves only as an initial, approximate estimate for $K_{c\,ini}$.

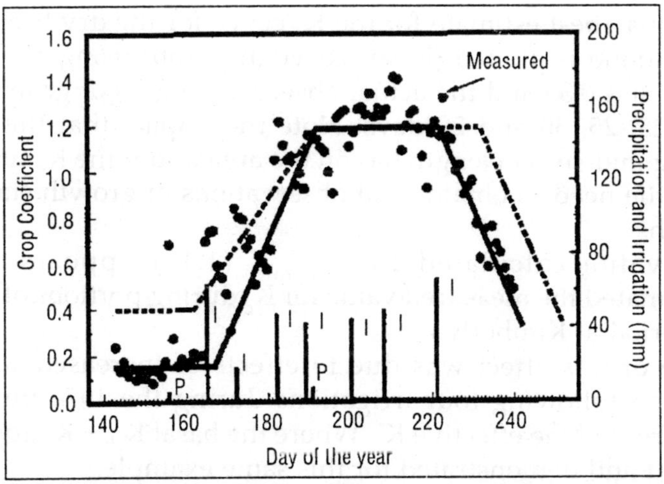

Constructed K_c curves using values from Tables directly (dotted line) and modified using $K_{c\,ini}$ from and L_{ini} = 25, L_{dev} = 25, L_{mid} = 30, and L_{late} = 20 days (heavy line) for dry beans at Kimberly, Idaho. Also shown are daily measured K_c.

$K_{c\,ini}$ can be more accurately estimated using the approach described in this chapter. ET_o during the initial period at Kimberly (late May - early June, 1974) averaged 5.3 mm/day, and the wetting interval during this period was approximately 14 days (2 rainfall events occurred averaging 5 mm per event).

Therefore, as the wetting events were light (< 10 mm each). The soil texture at Kimberly, Idaho is silt loam. $K_{c\,ini}$ for the 14 day wetting interval and ET_o = 5.3 mm/day is about 0.15. This value is substantially less than the general 0.4 value suggested, and emphasizes the need to utilize local, actual precipitation and irrigation data when determining $K_{c\,ini}$.

Comparison of Constructed Curves with Measurements

Because the ET_c data for the dry bean crop at Kimberly, Idaho were measured using a precision lysimeter system during 1974 by Wright, the actual K_c measurements can be compared with the constructed K_c curves, where actual K_c was calculated by dividing lysimeter measurements of ET_c by daily ET_o estimated using the FAO Penman-Monteith equation. As illustrated in the graph, the mid-season length for the general, continental climate overestimated the true mid-season length for dry beans in southern Idaho, which averaged only about 30 days rather than 40 days. This illustrates the importance of using the local observation of 30 days for mid-season period length rather than the general value.

The final, best estimate for the K_c curve for the dry bean crop in southern Idaho is plotted (lower curve in graph) using K_c values of 0.15, 1.19, and 0.35 and the actual observed lengths of growth stages equal to 25, 25, 30 and 20 days. Note the impact that the error in estimating mid-season length has on the area under the K_c curve. This supports the need to obtain local observations of growth stage dates and lengths.

The value calculated for $K_{c\ mid}$ (1.19) appears to have underestimated the measured value for K_c during portions of the mid-season period at Kimberly.

Some of this effect was due to effects of increased soil water evaporation following four irrigations during the 1974 mid-season which increased the effective K_c. Where the basal $K_{cb} + K_e$ approach is introduced and demonstrated for this same example.

The 0.15 value calculated for $K_{c\ ini}$ agrees closely with measured K_c during the initial period. Measured K_c during the development period exceeded the final K_c curve during days on or following wetting events. The day to day variation in the lysimeter measured K_c is normal and is caused by day to day variations in weather, in wind direction, by errors in prediction of R_n and ET_o, and by some random errors in the lysimeter measurements and weather measurements.

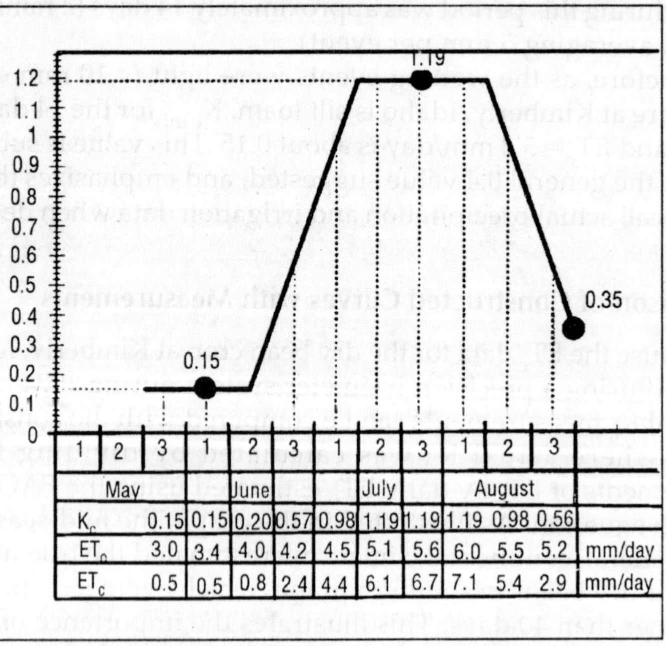

	May 1	May 2	May 3	June 1	June 2	June 3	July 1	July 2	July 3	Aug 1	Aug 2	Aug 3	
K_c			0.15	0.15	0.20	0.57	0.98	1.19	1.19	1.19	0.98	0.56	
ET_o			3.0	3.4	4.0	4.2	4.5	5.1	5.6	6.0	5.5	5.2	mm/day
ET_c			0.5	0.5	0.8	2.4	4.4	6.1	6.7	7.1	5.4	2.9	mm/day

Fig. K_c Curve and Ten-day Values for K_c and ET_c Derived from the Graph for the Drv Bean Crop

Example. First five days of that decade, $K_c = 0.15$, while during the second part of the decade K_c varies from 0.15 to 0.36 at the end of day 10. The K_c for that decade is consequently: 5/10 (0.15) + 5/10(0.15 + 0.36)/2 = 0.20.

Numerical Determination of K_c

The K_c coefficient for any period of the growing season can be derived by considering that during the initial and mid-season stages K_c is constant and equal to the K_c value of the growth stage under consideration. During the crop development and late season stage, K_c varies linearly between the K_c at the end of the previous stage ($K_{c\,prev}$) and the K_c at the beginning of the next stage ($K_{c\,next}$), which is $K_{c\,end}$ in the case of the late season stage:

$$K_{ci} = K_{cprev} + \left[\frac{i - \Sigma(L_{prev})}{L_{stage}}\right](K_{cnext} - K_{cprev})$$

where, i day number within the growing season [1.. length of the growing season],

$K_{c\,i}$ crop coefficient on day i,

L_{stage} length of the stage under consideration [days],

Σ (L_{prev}) sum of the lengths of all previous stages [days].

Example. Numerical determination of K_c

Determine K_c at day 20, 40, 70 and 95 for the dry bean crop (Figure 36).		
Crop growth stage	Length (days)	K_c
initial	25	$K_{c\,ini}$ = 0.15
crop development	25	0.15... 1.19
mid-season	30	$K_{c\,mid}$ = 1.19
late season	20	1.19 $K_{c\,end}$ = 0.35

At i = 20:	Initial stage, $K_c = K_{c\,ini}$ =	0.15	-
At i = 40	Crop development stage,		
For:	Σ (L_{prev}) = L_{ini} =	25	days
and:	$L_{stage} = L_{dev}$ =	25	days
From Equation	K_c = 0.15 + [(40 - 25)/25](1.19 - 0.15) =	0.77	-
At i = 70:	Mid-season stage, $K_c = K_{c\,mid}$ =	1.19	-
At i = 95	Late season stage,		
For:	Σ (L_{prev}) = $L_{ini} + L_{dev} + L_{mid}$ = (25 + 25 + 30) =	80	days
and:	$L_{stage} = L_{late}$ =	20	days
From Equation	K_c = 1.19 + [(95-80)/20](0.35-1.19) =	0.56	-
The crop coefficients at day 20, 40, 70 and 95 for the dry bean crop are 0.15, 0.77, 1.19 and 0.56 respectively.			

Alfalfa-Based Crop Coefficients

As two reference crop definitions (grass and alfalfa) are in use in various parts of the world, two families of K_c curves for agricultural crops have been developed. These are the alfalfa-based K_c curves by Wright and grass-based curves by Pruitt and those reported in this paper.

The user must exercise caution to avoid mixing grass-based K_c values with alfalfa reference ET and vice versa. Usually, a K_c based on the alfalfa reference can be 'converted' for use with a grass reference by multiplying by a factor ranging from about 1.0 to 1.3, depending on the climate (1.05 for humid, calm conditions, and 1.2 for semi-arid, moderately windy conditions, and 1.35 for arid, windy conditions):

$$K_{c\,(grass)} = K_{ratio}\,K_{c\,(alfalfa)}$$

where, $K_{c\,(grass)}$ grass-based K_c (this handbook),

$K_{c\,(alfalfa)}$ alfalfa-based K_c,

K_{ratio} conversion factor (1.0... 1.3).

A reference conversion ratio can be established for any climate by using the $K_{c\,mid}$ = 1.20 listed for alfalfa and then adjusting this $K_{c\,mid}$ for the climate using Equation. For example, at Kimberly, Idaho, the United States, where RH_{min} = 30% and u_2 = 2.2 m/s are average values during the summer months, a reference conversion ratio between alfalfa and grass references using Equation is approximately:

$$K_{ratio} = 1.2 + [0.04(2.2-2) - 0.004(30-45)]\left(\frac{0.5}{3}\right)^{0.3} = 1.24$$

where, h = 0.5 m is the standard height for the alfalfa reference.

Transferability of Previous K_c Values

The values for $K_{c\,mid}$ and $K_{c\,end}$ listed in Table are for a large part based on the original values presented in FAO Irrigation and Drainage, with some adjustment and revisions to reflect recent findings. Similarly adjustments in $K_{c\,mid}$ to compensate for differences in aerodynamic roughness and leaf area, as introduced in Equation are derived from the K_c values given for different wind and RH_{min} conditions in the concerned K_c Table, with some upward adjustment to better reflect increased ET_{crop} values under high wind and low RH_{min} when applied with the FAO Penman-Monteith equation.

The K_c's from FAO-24 were based primarily on a living grass reference crop. The FAO Penman-Monteith equation presented in this publication similarly represents the same standardized grass reference. For that reason K_c values are in general not very different between these publications except under high wind and low RH_{min}.

The modified Penman was found frequently to overestimate ET_o even up to 25% under high wind and low evaporative conditions and required often substantial local calibration. K_c values derived from crop water use studies which used the FAO-24 Penman equation to compute grass reference crop evapotranspiration, can therefore not be used and need to be adjusted using ET_o values estimated from the FAO Penman-Monteith equation.

Similarly crop water requirement estimates based on the FAO-24 Modified Penman equation will need to be reassessed in view of the found differences between the FAO-24 Penman and the FAO Penman-Monteith reference equations.

Chapter 6

Dual Crop Coefficient

This chapter also deals with the calculation of crop evapotranspiration (ET_c) under standard conditions where no limitations are placed on crop growth or evapotranspiration. This chapter presents the procedure for predicting the effects of specific wetting events on the value for the crop coefficient K_c. The solution consists of splitting K_c into two separate coefficients, one for crop transpiration, i.e., the basal crop coefficient (K_{cb}), and one for soil evaporation (K_e):

$$ET_c = (K_{cb} + K_e)\, ET_o$$

The dual crop coefficient approach is more complicated and more computationally intensive than the single crop coefficient approach (K_c). The procedure is conducted on a daily basis and is intended for applications using computers. It is recommended that the approach be followed when improved estimates for K_c are needed, for example to schedule irrigations for individual fields on a daily basis.

The calculation procedure for crop evapotranspiration, ET_c, consists of:

- Identifying the lengths of crop growth stages, and 'selecting the corresponding K_{cb} coefficients.
- Adjusting the selected K_{cb} coefficients for climatic conditions during the stage.
- Constructing the basal crop coefficient curve (allowing one to determine K_{cb} values for any period during the growing period).
- Determining daily K_e values for surface evaporation.
- Calculating ET_c as the product of ET_o and (K_{cb} + K_e).

TRANSPIRATION COMPONENT (K_{CB} ET_O)

Basal Crop Coefficient (K_{cb})

The basal crop coefficient (K_{cb}) is defined as the ratio of the crop evapotranspiration over the reference evapotranspiration (ET_c/ET_o)

when the soil surface is dry but transpiration is occurring at a potential rate, i.e., water is not limiting transpiration. Therefore, 'K_{cb} ET_o' represents primarily the transpiration component of ET_c. The K_{cb} ET_o does include a residual diffusive evaporation component supplied by soil water below the dry surface and by soil water from beneath dense vegetation.

As the K_c values include averaged effects of evaporation from the soil surface, the K_{cb} values lie below the K_c values and a separate Table for K_{cb} is required. Recommended values for K_{cb} for the same crops. The values for K_{cb} in the Table represent K_{cb} for a sub-humid climate and with moderate wind speed. For specific adjustment in climates where RH_{min} differs from 45% or where the wind speed is larger or smaller than 2 m/s, the $K_{cb\ mid}$ and $K_{cb\ end}$ values larger than 0.45 must be adjusted using the following equation:

$$K_{cb} = K_{cb(Tab)} + [0.04(u_2 - 2) - 0.004(RH_{min} - 45)]\left(\frac{h}{3}\right)^{0.3}$$

where, $K_{cb\ (Tab)}$ the value for $K_{cb\ mid}$ or $K_{cb\ end}$ (if ≥ 0.45) taken from table,

u_2 the mean value for daily wind speed at 2 m height over grass during the mid or late season growth stage [m s^{-1}] for 1 m $s^{-1} \leq u_2 \leq 6$ m s^{-1},

RH_{min} the mean value for daily minimum relative humidity during the mid- or late season growth stage [%] for $20\% \leq RH_{min} \leq 80\%$,

h the mean plant height during the mid or late season stage [m] for $20\% \leq RH_{min} \leq 80\%$.

For a full discussion on the impact of the climatic correction, and the numerical determination of $K_{cb\ mid}$ and $K_{cb\ end}$, the user is referred to the discussions on $K_{c\ mid}$ and $K_{c\ end}$.

The general guidelines that were used in deriving K_{cb} values from the K_c values. Where local research results are available, values for K_{cb} from Table can be modified to reflect effects of local conditions, cultural practices or crop varieties on K_{cb}.

However, local values for K_{cb} should not be expected to deviate by more than 0.2 from the values.

A greater deviation should signal the need to investigate or evaluate the local research technique, equipment and cultural practices. Where local K_{cb} values are used, no adjustment for climate using Equation 70 is necessary.

Selection and Adjustment of Basal Crop Coefficients, K_{cb}

Select $K_{cb\ ini}$, $K_{cb\ mid}$ and $K_{cb\ end}$ for the dry bean crop.

$K_{cb\ ini}$, $K_{cb\ mid}$ and $K_{cb\ end}$ can be selected directly from table for dry beans as 0.15, 1.10 and 0.25. When adjusted for climate using Equation:

$K_{cb\ ini} = 0.15$

$K_{cb\ mid} = 1.10 + (0.04\ (2.2\text{-}2) - 0.004\ (30\text{-}45))(0.4/3)^{0.3} = 1.14$

$K_{cb\ end} = 0.25$ (as $K_{cb} < 0.45$)

Table. Basal Crop Coefficients, K_c, for Mon Stressed, Well-managed Crops in Subhumid Climates ($RH_{min} \approx 45\%$, $u_2 \approx 2$ m/s) for Use with the FAO Penman-Monteith ET_o.

Crop	$K_{cb\,ini}$[1]	$K_{cb\,mid}$	$K_{cb\,end}$
a. Small Vegetables	0.15	0.95	0.85
Broccoli		0.95	0.85
Brussel Sprouts		0.95	0.85
Cabbage		0.95	0.85
Carrots		0.95	0.85
Cauliflower		0.95	0.85
Celery		0.95	0.90
Garlic		0.90	0.60
Lettuce		0.90	0.90
Onions			
- dry		0.95	0.65
- green		0.90	0.90
- seed		1.05	0.70
Spinach		0.90	0.85
Radishes		0.85	0.75
b. Vegetables - Solanum Family *(Solanaceae)*	0.15	1.10	0.70
Egg Plant		1.00	0.80
Sweet Peppers (bell)		1.00[2]	0.80
Tomato		1.10[2]	0.60-0.80
c. Vegetables - Cucumber Family *(Cucurbitaceae)*	0.15	0.95	0.70
Cantaloupe		0.75	0.50
Cucumber			
- Fresh Market		0.95[2]	0.70
- Machine harvest		0.95	0.80

Determination of Daily K_{cb} Values

Only three point values are required to describe and to construct the crop coefficient curve. After dividing the growing period into the four general growth stages and selecting and adjusting the K_{cb} values corresponding to the initial ($K_{cb\ ini}$), mid-season ($K_{cb\ mid}$) and end of the late season stages ($K_{cb\ end}$), the crop coefficient curve can be drawn and the K_{cb} coefficients can be derived.

Pumpkin, Winter Squash		0.95	0.70
Squash, Zucchini		0.90	0.70
Sweet Melons		1.00	0.70
Watermelon		0.95	0.70
d. Roots and Tubers	0.15	1.00	0.85
Beets, table		0.95	0.85
Cassava			
- year 1		0.70[3]	0.20
- year 2		1.00	0.45
Parsnip		0.95	0.85
Potato		1.10	0.65[4]
Sweet Potato		1.10	0.55
Turnip (and Rutabaga)		1.00	0.85
Sugar Beet		1.15	0.50[5]
c. Lugumes *(Leguminosae)*	0.15	1.10	0.50
Beans, green		1.00[2]	0.80
Beans, dry and Pulses		1.10[2]	0.25
Chick pea		0.95	0.25
Fababean (broad bean)			
- Fresh		1.10[2]	1.05
- Dry/Seed		1.10[2]	0.20
Grabanzo		1.05	0.25
Green Gram and Cowpeas		1.00	0.55-0.25[6]
Groundnut (Peanut)		1.10	0.50
Lentil		1.05	0.20
Peas			
- Fresh		1.10[2]	1.05
- Dry/Seed		1.10	0.20
Soybeans		1.10	0.30
f. Perennial Vegetables (with winter dormancy and initially bare or mulched soil)			
Artichokes	0.15	0.95	0.90
Asparagus	0.15	0.90[7]	0.20
Mint	0.40	1.10	1.05
Strawberries	0.30	0.80	0.70
g. Fibre Crops	0.15		
Cotton		1.10-1.15	0.50-0.40
Flax		1.05	0.20
Sisal [8]		0.4-0.7	0.4-0.7
h. Oil Crops	0.15	1.10	0.25
Castorbean *(Ricinus)*		1.10	0.45
Rapeseed, Canola		0.95-1.10[9]	0.25
Safflower		0.95-1.10[9]	0.20
Sesame		1.05	0.20
Sunflower		0.95-1.10[9]	0.25
i. Cereals	0.15	1.10	0.25
Barley		1.10	0.15
Oats		1.10	0.15

Spring Wheat		1.10	0.15-0.3[10]
Winter Wheat	0.15-0.5[11]	1.10	0.15-0.3[10]
Maize			
- Field (grain) *(field corn)*	0.15	1.15	0.50,0.15[12]
- Sweet *(sweet corn)*		1.10	1.00[13]
Millet		0.95	0.20
Sorghum			
- grain		0.95-1.05	0.35
- sweet		1.15	1.00
Rice	1.00	1.15	0.70-0.45
j. Forages			
Alfalfa Hay			
- individual cutting periods	0.30[14]	1.15[14]	1.10[14]
- for seed	0.30	0.45	0.45
Bermuda hay			
- averaged cutting effects	0.50	0.95[15]	0.80
- Spring crop for seed	0.15	0.85	0.60
Clover hay, Berseem - individual cutting periods	0.30[14]	1.10[14]	1.05[14]
Rye Grass hay - averaged cutting effects	0.85	1.00[15]	0.95
Sudan Grass hay (annual) - individual cutting periods	0.30[14]	1.10[14]	1.05[14]
Grazing Pasture			
- Rotated Grazing	0.30	0.80-1.00	0.80
- Extensive Grazing	0.30	0.70	0.70
Turf grass			
- cool season [16]	0.85	0.90	0.90
- warm season [16]	0.75	0.80	0.80
k. Sugar cane	0.15	1.20	0.70
l. Tropical Fruits and Trees			
Banana			
- 1st year	0.15	1.05	0.90
- 2nd year	0.60	1.10	1.05
Cacao	0.90	1.00	1.00
Coffee			
- bare ground cover	0.80	0.90	0.90
- with weeds	1.00	1.05	1.05
Date Palms	0.80	0.85	0.85
Palm Trees	0.85	0.90	0.90
Pineapple [17] (multiyear crop)			
- bare soil	0.15	0.25	0.25
- with grass cover	0.30	0.45	0.45
Rubber Trees	0.85	0.90	0.90
Tea			
- nonshaded	0.90	0.95	0.90
- shaded [18]	1.00	1.10	1.05

m. Grapes and Berries			
Berries (bushes)	0.20	1.00	0.40
Grapes			
- Table or Raisin	0.15	0.80	0.40
- Wine	0.15	0.65	0.40
Hops	0.15	1.00	0.80
n. Fruit Trees			
Almonds, no ground cover	0.20	0.85	0.6019
Apples, Cherries, Pears [20]			
- no ground cover, killing frost	0.35	0.90	0.6519
- no ground cover, no frosts	0.50	0.90	0.7019
- active ground cover, killing frost	0.45	1.15	0.9019
- active ground cover, no frosts	0.75	1.15	0.8019
Apricots, Peaches, Stone Fruit [20, 21]			
- no ground cover, killing frost	0.35	0.85	0.6019
- no ground cover, no frosts	0.45	0.85	0.6019
- active ground cover, killing frost	0.45	1.10	0.8519
- active ground cover, no frosts	0.75	1.10	0.8019
Avocado, no ground cover	0.50	0.80	0.70
Citrus, no ground cover [22]			
70% canopy	0.65	0.60	0.65
50% canopy	0.60	0.55	0.60
20% canopy	0.45	0.40	0.50
Citrus, with active ground cover or weeds [23]			
70% canopy	0.75	0.70	0,75
50% canopy	0.75	0.75	0.75
20% canopy	0.80	0.80	0.85
Conifer Trees [24]	0.95	0.95	0.95
Kiwi	0.20	1.00	1.00
Olives (40 to 60% ground coverage by canopy) [25]	0.55	0.65	0.65
Pistachios, no ground cover	0.20	1.05	0.40
Walnut Orchard [20]	0.40	1.05	0.6019

Determination of Daily Values for K_{cb}

Calculate the basal crop coefficient for the dry beans at the middle of each of the four growth stages.

Initial stage (L_{ini} = 25 days), at day 12 of the growing period:

$$K_{cb} = K_{cb\ ini} = 0.15$$

Crop development stage (L_{dev} = 25 days), at day (25 + 25/2 =) 37 of the growing period, using Equation:

$$K_{cb} = 0.15 + [(37 - 25)/25]\ (1.14 - 0.15) = 0.63$$

Table. General Guidelines to Derive K_{cb} from the K_c Values Listed in Table

Growth stage	Ground condition, irrigation and cultural practices	K_{cb}	further adjustment
Initial	Annual crop - (nearly) bare soil surface.	0.15	
	Perennial crop - (nearly) bare soil surface	0.15-0.20	-
	Grasses, brush and trees - killing frost	0.30 - 0.40	-
	Perennial crop - some ground cover or leaf cover		
	- infrequently irrigated (olives, palm trees, fruit trees,...)	$K_{c\ ini\ (Tab.12)}$ - 0.1	-
	- frequently irrigated (garden-type vegetables,...)	$K_{c\ ini\ (Tab.12)}$ - 0.2	-
Mid-season	Ground cover more than 80%	$K_{c\ mid\ (Tab.12)}$ - 0.05	Climate (Eq. 70)
	Ground cover less than 80% (vegetables)	$K_{c\ mid\ (Tab.12)}$ - 0.10	Climate (Eq. 70)
At end of season	infrequently irrigated or wetted during late season	~ $K_{c\ end}$ - 0.05	-
	frequently irrigated or wetted during late season	$K_{c\ end}$ - 0.1	Climate (Eq. 70)

Mid-season stage (L_{mid} = 30 days), at day (25 + 25 + 30/2 =) 65 of the growing period:

$$K_{cb} = K_{cb\ mid} = 1.14$$

Late season stage (L_{late} = 20 days), at day (25 + 25 + 30 + 20/2 =) 90 of the growing period, Equation :

$$K_{cb} = 1.14 + [(90 - (25 + 25 + 30))/20]\ (0.25 - 1.14) = 0.70$$

The basal crop coefficients, K_{cb}, at days 12, 37, 65 and 90 of the growing period are 0.15, 0.63, 1.14 and 0.70 respectively.

Evaporation Component (K_e ET_o)

The soil evaporation coefficient, K_e, describes the evaporation component of ET_c. Where the topsoil is wet, following rain or irrigation, K_e is maximal. Where the soil surface is dry, K_e is small and even zero when no water remains near the soil surface for evaporation.

Calculation Procedure

Where the soil is wet, evaporation from the soil occurs at the maximum rate. However, the crop coefficient ($K_c = K_{cb} + K_e$) can never exceed a maximum value, $K_{c\ max}$. This value is determined by the energy available for evapotranspiration at the soil surface

$$(K_{cb} + K_e \leq K_{c\ max}) \text{ or } K_e \leq (K_{c\ max} - K_{cb}).$$

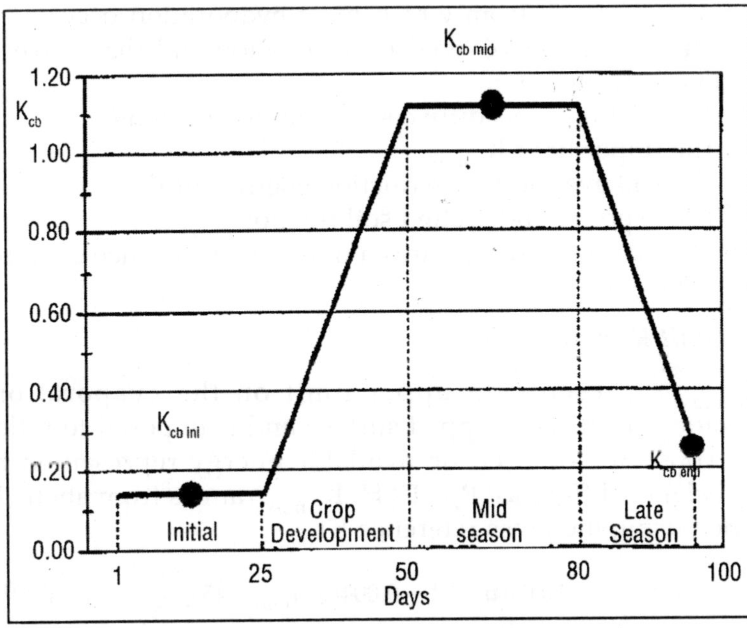

Fig. Constructed Basal Crop Coefficient (K_{cb}) Curve for a Dry Bean Crop using Growth Stage Lengths of 25, 25, 30 and 20 Days.

When the topsoil dries out, less water is available for evaporation and a reduction in evaporation begins to occur in proportion to the amount of water remaining in the surface soil layer, or:

$$K_e = K_r (K_{c\ max} - K_{cb}) \leq f_{ew} K_{c\ max}$$

where

K_e soil evaporation coefficient,

K_{cb} basal crop coefficient,

$K_{c\ max}$ maximum value of K_c following rain or irrigation,

K_r dimensionless evaporation reduction coefficient dependent on the cumulative depth of water depleted (evaporated) from the topsoil,

f_{ew} fraction of the soil that is both exposed and wetted, i.e., the fraction of soil surface from which most evaporation occurs.

In computer programming terminology, Equation is expressed as

$$K_e = \min (K_r (K_{c\ max} - K_{cb}), f_{ew} K_{c\ max}).$$

Following rain or irrigation K_r is 1, and evaporation is only determined by the energy available for evaporation. As the soil surface dries, K_r becomes less than one and evaporation is reduced. K_r becomes zero when no water is left for evaporation in the upper soil layer.

Evaporation occurs predominantly from the exposed soil fraction. Hence, evaporation is restricted at any moment by the energy available at the exposed soil fraction, i.e., K_e cannot exceed $f_{ew}\ K_{c\,max}$, where f_{ew} is the fraction of soil from which most evaporation occurs, i.e., the fraction of the soil not covered by vegetation and that is wetted by irrigation or precipitation.

The calculation procedure consists in determining:

- The upper limit $K_{c\,max}$.
- The soil evaporation reduction coefficient K_r.
- The exposed and wetted soil fraction f_{ew}.

The estimation of K_r requires a daily water balance computation for the surface soil layer.

Upper Limit K_c max

$K_{c\,max}$ represents an upper limit on the evaporation and transpiration from any cropped surface and is imposed to reflect the natural constraints placed on available energy represented by the energy balance difference R_n - Γ- H. $K_{c\,max}$ ranges from about 1.05 to 1.30 when using the grass reference ET_o:

$$K_{c\,max} = \max\left(\left\{1.2 + [0.04(u_2 - 2) - 0.004(RH_{min} - 45)]\left(\frac{h}{3}\right)^{0.3}\right\}\{K_{cb} + 0.05\}\right)$$

where, h mean maximum plant height during the period of calculation (initial, development, mid-season, or late-season),

K_{cb} basal crop coefficient,

max ( ) maximum value of the parameters in braces {} that are separated by the comma.

Equation ensures that $K_{c\,max}$ is always greater or equal to the sum K_{cb} + 0.05. This requirement suggests that wet soil will always increase the value for K_{cb} by 0.05 following complete wetting of the soil surface, even during periods of full ground cover.

A value of 1.2 instead of 1 is used for $K_{c\,max}$ in Equation because of the effect of increased aerodynamic roughness of surrounding crops during development, mid-season and late season growth stages which can increase the turbulent transfer of vapour from the exposed soil surface. The "1.2" coefficient also reflects the impact of the reduced albedo of wet soil and the contribution of heat stored in dry soil prior to the wetting event. All of these factors can contribute to increased evaporation relative to the reference.

The "1.2" coefficient in Equation represents effects of wetting intervals that are greater than 3 or 4 days. If irrigation or precipitation events are more frequent, for example daily or each two days, then the soil has less opportunity to absorb heat between wettings, and the

"1.2" coefficient in Equation can be reduced to about 1.1. The time step to compute $K_{c\,max}$ may vary from daily to monthly.

Soil Evaporation Reduction Coefficient (K_r)

Soil evaporation from the exposed soil can be assumed to take place in two stages: an energy limiting stage, and a falling rate stage. When the soil surface is wet, K_r is 1. When the water content in the upper soil becomes limiting, K_r decreases and becomes zero when the total amount of water that can be evaporated from the topsoil is depleted.

Maximum Amount of Water that can be Evaporated

In the simple evaporation procedure, it is assumed that the water content of me evaporating layer of the soil is at field capacity, θ_{FC} shortly following a major wetting event and that the soil can dry to a soil water content level that is halfway between oven dry (no water left) and wilting point, θ_{WP}. The amount of water that can be depleted by evaporation during a complete drying cycle can hence be estimated as:

$$TEW = 1000\,(\theta_{FC} - 0.5\,\theta_{WP})\,Z_e$$

where, TEW total evaporable water = maximum depth of water that can be evaporated from the soil when the topsoil has been initially completely wetted [mm],

θ_{FC} soil water content at field capacity [$m^3\,m^{-3}$],

θ_{WP} soil water content at wilting point [$m^3\,m^{-3}$],

Z_e depth of the surface soil layer that is subject to drying by way of evaporation [0.10-0.15m].

Where unknown, a value for Z_e, the effective depth of the soil evaporation layer, of 0.10-0.15 m is recommended. Typical values for θ_{FC}, θ_{WP} and TEW are given in table.

Typical Soil Water Characteristics for Different Soil Types

Stage 1: Energy Limiting Stage

At the start of a drying cycle, following heavy rain or irrigation, the soil water content in the topsoil is at field capacity and the amount of water depleted by evaporation, D_e, is zero. During stage 1 of the drying process, the soil surface remains wet and it is assumed that evaporation from soil exposed to the atmosphere will occur at the maximum rate limited only by energy availability at the soil surface.

This stage holds until the cumulative depth of evaporation, D_e, is such that the hydraulic properties of the upper soil become limiting and water cannot be transported to the soil surface at a rate that can supply the potential demand. During stage 1 drying, $K_r = 1$.

Soil type (USA Soil Texture Classification)	Soil water characteristics			Evaporation parameters	
	θ FC	θ WP	(θ FC - θ WP)	Amount of water that can be depleted by evaporation	
				stage 1 REW	stages 1 and 2 TEW* (Z_e = 0.10m)
	m^3/m^3	m^3/m^3	m^3/m^3	mm	mm
Sand	0.07 - 0.17	0.02 - 0.07	0.05 - 0.11	2 - 7	6 - 12
Loamy sand	0.11 - 0.19	0.03 - 0.10	0.06 - 0.12	4 - 8	9 - 14
Sandy loam	0.18 - 0.28	0.06 - 0.16	0.11 - 0.15	6 - 10	15 - 20
Loam	0.20 - 0.30	0.07 - 0.17	0.13 - 0.18	8 - 10	16 - 22
Silt loam	0.22 - 0.36	0.09 - 0.21	0.13 - 0.19	8 - 11	18 - 25
Silt	0.28 - 0.36	0.12 - 0.22	0.16 - 0.20	8 - 11	22 - 26
Silt clay loam	0.30 - 0.37	0.17 - 0.24	0.13 - 0.18	8 - 11	22 - 27
Silty clay	0-30 - 0.42	0.17 - 0.29	0.13 - 0.19	8 - 12	22 - 28
Clay	0.32 - 0.40	0.20 - 0.24	0.12 - 0.20	8 - 12	22 - 29

Fig. Soil Evaporation Reduction Coefficient, K_r

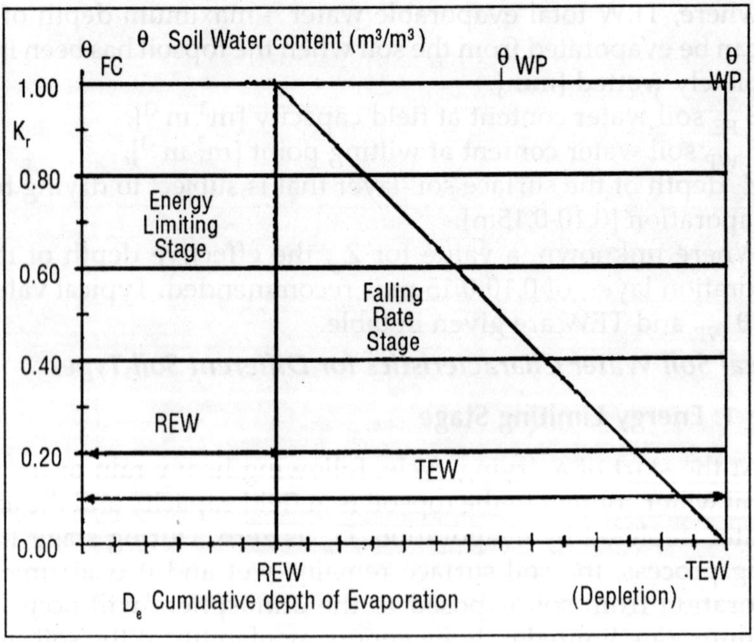

The cumulative depth of evaporation, De, at the end of stage 1 drying is REW (Readily evaporable water, which is the maximum depth of water that can be evaporated from the topsoil layer without restriction during stage 1). The depth normally ranges from 5 to 12 mm and is generally highest for medium and fine textured soils.

Stage 2: Falling Rate Stage

The second stage (where the evaporation rate is reducing) is termed the 'falling rate stage' evaporation and starts when D_e exceeds REW. At this point, the soil surface is visibly dry, and the evaporation from the exposed soil decreases in proportion to the amount of water remaining in the surface soil layer:

$$K_r = \frac{TEW - D_{ei-1}}{TEW - REW} \text{ for} D_{ei-1} > REW$$

where, K_r dimensionless evaporation reduction coefficient dependent on the soil water depletion (cumulative depth of evaporation) from the topsoil layer ($K_r = 1$ when $D_{e,\,i-1} \leq REW$),

$D_{e,\,i-1}$ cumulative depth of evaporation (depletion) from the soil surface layer at the end of day $_{i-1}$ (the previous day) [mm],

TEW maximum cumulative depth of evaporation (depletion) from the soil surface layer when $K_r = 0$ (TEW = total evaporable water) [mm],

REW cumulative depth of evaporation (depletion) at the end of stage 1 (REW = readily evaporable water) [mm].

Determination of the Evapotranspiration from a Bare Soil

Determine the evapotranspiration from a bare loamy soil surface ($K_{cb} \approx 0.15$) for ten successive days following a heavy rain. The reference evapotranspiration during the drying period is ET_o = 4.5 mm/day, and the climate is subhumid with light wind.							
From Table 19			For Loam: TEW ≈ 20 mm and REW ≈ 9 mm				
For rain on bare soil			f_{ew} = 1				
From Eq. 72			$K_{c\,max}$ = 1.20				
(1)	(2)	(3)	(4)	(5)	(6)	(7)	(8)
Day	D_e start mm	Stage	K_r	K_e	$K_e\ ET_o$ mm/day	$D_{e\ end}$ mm	ET_c mm/day
1	0.00	1	1	1.05	4.73	4.73	5.4
2	4.73	1	1	1.05	4.73	9.45	5.4
3	9.45	2	(20 - 9.45)/(20 - 9) = 0.96	1.01	4.53	13.98	5.2
4	13.98	2	(20 - 13.98)/(20 - 9) = 0.55	0.57	2.59	16.57	3.3
5	16.57	2	(20 - 16.57)/(20 - 9) = 0.31	0.33	1.47	18.04	2.1
6	18.04	2	(20 - 18.04)/(20 - 9) = 0.18	0.19	0.84	18.88	1.5

7	18.88	2	(20 - 18.88/(20 - 9) = 0.10	0.11	0.48	19.36	1.2
8	19.36	2	(20 - 19.36)/(20 - 9) = 0.06	0.06	0.27	19.64	0.9
9	19.64	2	(20 - 19.64)/(20 - 9) = 0.03	0.03	0.16	19.79	0.8
10	19.79	2	(20 - 19.79)/(20 - 9) = 0.02	0.02	0.09	19.88	0.8

(1)	Day number.
(2)	Depletion at beginning of the day (= depletion at end of previous day).
(3)	Soil evaporation stage (stage 2 starts if D_e > REW = 9 mm).
(4)	K_r (K_r = 1 for stage 1. Use Eq. 74 for stage 2).
(5)	From Eq. 21: $K_e = K_r (K_{c\,max} - K_{cb}) = K_r (1.20\text{-}0.15) = 1.05\, K_r \leq 1.20$.
(6)	Evaporation component: $K_e\, ET_o = K_e$ (4.5 mm/day).
(7)	Depletion at end of day = (2) - (6).
(8)	$ET_c = (K_{cb} + K_e)\, ET_o = (0.15 + K_e)\, ET_o = (0.15 + K_e)$ 4.5 mm/day, where $K_{cb}\, ET_o = (0.15\, ET_o) \approx 0.7$ mm/day is basal, "diffusive" evaporation from the soil, possibly from beneath the Z_e depth (~ 0.10 to 0.15 m). Since the soil in this situation is bare, one could set the K_{cb} equal to zero so that maximum K_e becomes $K_e = K_{c\,max} = 1.20$. Then all of the evaporation would be deducted from the surface soil layer.

The example demonstrates that the estimation of K_r requires a daily water balance calculation. This is further developed in the section on the daily calculation of K_e.

Exposed and Wetted Soil Fraction (f_{ew})

Few: Calculation Procedure

In crops with incomplete ground cover, evaporation from the soil often does not occur uniformly over the entire surface, but is greater between plants where exposure to sunlight occurs and where more air ventilation is able to transport vapour from the soil surface to above the canopy. This is especially true where only part of the soil surface is wetted by irrigation.

It is recognized that both the location and the fraction of the soil surface exposed to sunlight change to some degree with the time of day and depending on row orientation. The procedure presented here predicts a general averaged fraction of the soil surface from which the majority of evaporation occurs. Diffusive evaporation from the soil beneath the crop canopy is assumed to be largely included in the basal K_{cb} coefficient.

Where the complete soil surface is wetted, as by precipitation or sprinkler, then the fraction of soil surface from which most evaporation occurs, f_{ew}, is essentially defined as $(1 - f_c)$, where f_c is the average fraction of soil surface covered by vegetation and is the approximate fraction of soil surface that is exposed. However, for irrigation systems where only a fraction of the ground surface is wetted, f_{ew} must be limited to f_w, the fraction of the soil surface wetted by irrigation. Therefore, f_{ew} is calculated as:

$$f_{ew} = \min(1 - f_c, f_w)$$

where, $1 - f_c$ average exposed soil fraction not covered (or shaded) by vegetation [0.01 - 1],

f_w average fraction of soil surface wetted by irrigation or precipitation [0.01 - 1].

The limitation imposed by Equation assumes that the fraction of soil wetted by irrigation occurs within the fraction of soil exposed to sunlight and ventilation. This is generally the case, except perhaps with drip irrigation.

In the case of drip irrigation, where the majority of soil wetted by irrigation may be beneath the canopy and may therefore be shaded, more complex models of the soil surface and wetting patterns may be required to accurately estimate total evaporation from the soil. In this case, the value for f_w may need to be reduced to about one-half to one-third of that given in Table to account for the effects of shading of emitters by the plant canopy on the evaporation rate from wetted soil. A general approach could be to multiply f_w by $[1 - (2/3)f_c]$ for drip irrigation.

fw: Fraction of Soil Surface Wetted by Irrigation or Precipitation

Table presents typical values for f_w. Where a mixture of irrigation and precipitation occur within the same drying period or on the same day, the value for f_w should be based on a weighted average of the f_w for precipitation ($f_w = 1$) and the f_w for the irrigation system. The weighting should be approximately proportional to the infiltration depths from each water source.

Alternatively, on each day of the application, the following rules can be applied to determine f_w for that and subsequent days in a more simplified manner:

- Surface is wetted by irrigation and rain: f_w is the f_w for the irrigation system.
- Surface is wetted by irrigation: f_w is the f_w for the irrigation system.
- Surface is wetted by significant rain (i.e., > 3 to 4 mm) with no irrigation: $f_w = 1$.

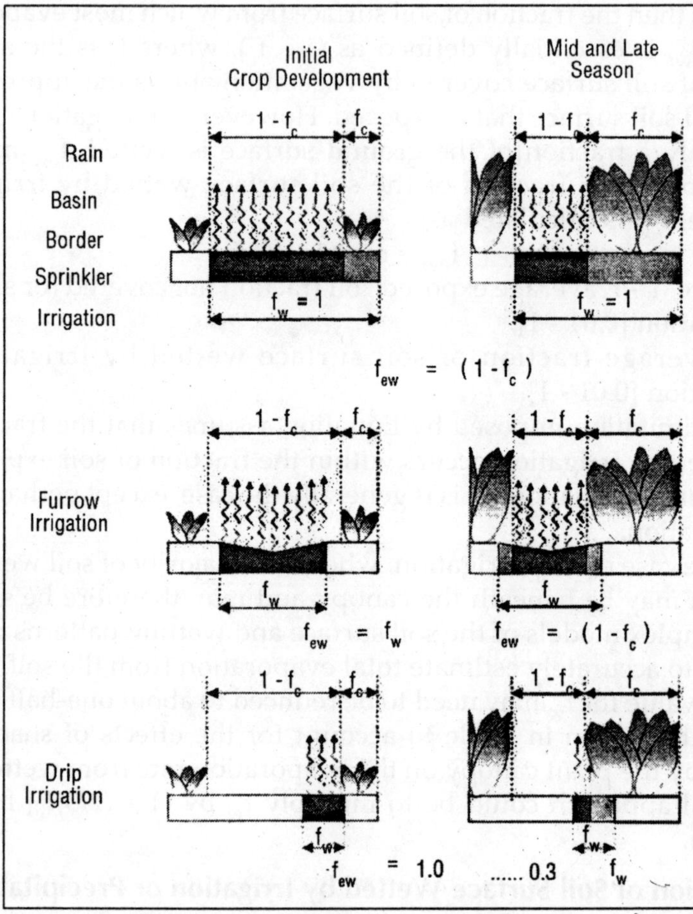

Fig. Determination of Variable f_{ew} (cross-hatched areas) as a Function of the Fraction of Ground Surface Coverage (f_c) and the Fraction of the Surface Wetted (f_w)

Table. Common Values of Fraction f_w of Soil Surface Wetted by Iirrigation or Precipitation

Wetting Event	f_w
Precipitation	1.0
Sprinkler irrigation	1.0
Basin irrigation	1.0
Border irrigation	1.0
Furrow irrigation (every furrow), narrow bed	0.6...1.0
Furrow irrigation (every furrow), wide bed	0.4... 0.6
Furrow irrigation (alternated furrows)	0.3...0.5
Trickle irrigation	0.3... 0.4

- Where there is neither irrigation nor significant precipitation: f_w is the f_w of the previous day.

1 – f_c: Exposed Soil Fraction

The fraction of the soil surface that is covered by vegetation is termed f_c. Therefore, $(1 - f_c)$ represents the fraction of the soil that is exposed to sunlight and air ventilation and which serves as the site for the majority of evaporation from wet soil.

The value for f_c is limited to < 0.99. The user should assume appropriate values for the various growth stages. Typical values for f_c and $(1 - f_c)$ are given in Table.

Table. Common Values of Fractions Covered by Vegetation (f_c) and Exposed to Sunlight ($1 - f_c$)

Crop growth stage	f_c	$1 - f_c$
Initial stage	0.0 - 0.1	1.0 - 0.9
Crop development stage	0.1 - 0.8	0.9 - 0.2
Mid-season stage	0.8 - 1.0	0.2 - 0.0
Late season stage	0.8 - 0.2	0.2 - 0.8

Where f_c is not measured, f_c can be estimated using the relationship:

$$f_c = \left(\frac{K_{cb} - K_{c\,min}}{K_{c\,max} - K_{c\,min}}\right)^{(1+0.5h)}$$

where, f_c the effective fraction of soil surface covered by vegetation [0 – 0.99],

K_{cb} the value for the basal crop coefficient for the particular day or period,

$K_{c\,min}$ the minimum K_c for dry bare soil with no ground cover [≈ 0.15 – 0.20],

$K_{c\,max}$ the maximum K_c immediately following wetting (Equation 72),

h mean plant height [m].

This equation should be used with caution and validated from field observations. $K_{c\,min}$ is the minimum crop coefficient for dry bare soil when transpiration and evaporation from the soil are near baseline (diffusive) levels. $K_{c\,min} \approx 0.15 - 0.20$ is recommended. The value of $K_{c\,min}$ is an integral part of all K_{cb} coefficients. $K_{c\,min}$ ordinarily has the same value as the $K_{cb\,ini}$ used for annual crops under nearly bare soil conditions (0.15 – 0.20).

Equation assumes that the value for K_{cb} is largely affected by the fraction of soil surface covered by vegetation. This is a good assumption for most vegetation and conditions.

The '1 + 0.5 h' exponent in the equation represents the effect of plant height on shading the soil surface and in increasing the value for K_{cb} given a specific value for f_c.

The user should limit the difference $K_{cb} - K_{c\,min}$ to ≥ 0.01 for numerical stability. The value for f_c will change daily as K_{cb} changes. Therefore, the above equation is applied daily.

Application of Equation predicts that f_c decreases during the late season period in proportion to K_{cb}, even though the ground may remain covered with senescing vegetation. This prediction helps to account for the local transport of sensible heat from senescing leaves to the soil surface below.

Calculation of the Crop Coefficient (Kcb + Ke) under Sprinkler Irrigation

A field of cotton has just been sprinkler irrigated. The K_{cb} for the specific day (during the development period) has been computed and then interpolated from the K_{cb} curve as 0.9. The ET_o = 7 mm/day, u_2 = 3 m/s and RH_{min} = 20%. Estimate the crop coefficient ($K_{cb} + K_e$).

Assuming h = 1 m, from Equation, $K_{c\,max}$ for this arid climate is:

$$K_{c\,max} = \max\left(\left\{1.2 + [0.04(3-2) - 0.004(20-45)]\left(\frac{1}{3}\right)^{0.3}\right\}\{0.9 + 0.05\}\right) = 1.30$$

From Equation , where $K_{c\,min}$ = 0.15:

$f_c = [(K_{cb} - K_{c\,min})/(K_{c\,max} - K_{c\,min})]^{(1+0.5h)}$
$= [(0.9 - 0.15)/(1.3 - 0.15)]^{(1+0.5(1))} = 0.53$.

As the field was sprinkler irrigated, f_w = 1.0 and from Equation:

$$f_{ew} = \min(1 - f_c, f_w) = \min(1 - 0.53, 1.0) = 0.47.$$

Assuming that the irrigation was sufficient to fill the evaporating layer to field capacity, so that K_r = 1, evaporation would be in stage 1.

From Equation: K_e = 1.00 (1.30 – 0.90) = 0.40

The value is compared against the upper limit $f_{ew}\,K_{c\,max}$ to ensure that it is less than the upper limit:

$f_{ew}\,K_{c\,max}$ = 0.47 (1.30) = 0.61, which is greater than the value for K_e. Therefore, the value for K_e can be used with no limitation.

The total K_c for the field, assuming no moisture stress due to a dry soil profile, is

$$K_c = K_{cb} + K_e = 0.9 + 0.40 = 1.30.$$

This value is large because of the very wet soil surface, the relatively tall rough crop as compared to the grass reference, and the arid climate (u_2 = 3 m/s and RH_{min} = 20%). In this situation, K_c happens to equal $K_{c\,max}$, as the field has just been wetted by sprinkler irrigation.

Calculation of the Crop Coefficient ($K_{cb} + K_e$) under Furrow Irrigation

The cotton field in the previous example has been irrigated by furrow irrigation of alternate rows rather than by sprinkler, and the fraction of the field surface wetted by the irrigation is 0.3.

The f_{ew} in this case is calculated from Equation as:

$$f_{ew} = \min(1 - f_c, f_w) = \min(1 - 0.53, 0.3) = 0.3.$$

Assuming that the irrigation was sufficient to fill the f_{ew} portion of the evaporating layer to field capacity, so that $K_r = 1$, evaporation would be in stage 1.

From Equation : $K_e = 1.00\ (1.30 - 0.9) = 0.40$

The value is compared to the upper limit $f_{ew}\ K_{c\,max}$ which is 0.30 (1.30) = 0.39. Because 0.40 > 0.39, K_e from the f_{ew} surface area is constrained to 0.39.

The total K_c for the furrow irrigated field, assuming no moisture stress due to dry soil, is $K_c = K_{cb} + K_e = 0.9 + 0.39 = 1.29$. This value is essentially the same as for the previous example because the procedure assumes that the soil between alternate rows is the portion that is wetted by the irrigation, so that the majority of the field surface has either vegetation cover or wet soil.

Calculation of the Crop Coefficient ($K_{cb} + K_e$) under Drip Irrigation

The cotton field in the previous example has been irrigated by drip irrigation, where the emitters are placed beneath the cotton canopy. The fraction of the field surface wetted by the irrigation is 0.3.

The f_{ew} in this case is calculated from Equation as $f_{ew} = \min(1 - f_c, f_w)$. Because the emitters are beneath the canopy so that less energy is available for evaporation, the value for f_w is reduced by multiplying by $1 - (2/3)f_c$, so that:

$$f_{ew} = \min[(1 - f_c),(1 - 0.67\ f_c)\ f_w)]$$
$$= \min[(1 - 0.53), (1 - 0.67(0.53))(0.3)] = 0.19$$

Assuming that the irrigation was sufficient to fill the f_w portion of the evaporating layer to field capacity, so that $K_r = 1$, evaporation would be in stage 1.

From Equation : $K_e = 1.00\ (1.30\text{-}0.90) = 0.40$.

The value is compared to the upper limit $f_{ew}\ K_{c\,max} = 0.19\ (1.30) = 0.25$. Because 0.25 < 0.40, K_e from the f_w fraction of the surface area is constrained by the available energy. Therefore $K_e = 0.25$.

The total K_c for the drip irrigated field, assuming no moisture stress due to dry soil, is:

$$K_c = K_{cb} + K_e = 0.9 + 0.25 = 1.15.$$

This K_c value is less than that for sprinkler and furrow irrigation.

DAILY CALCULATION OF K_E

Daily Water Balance

The estimation of K_e in the calculation procedure requires a daily water balance computation for the surface soil layer for the calculation of the cumulative evaporation or depletion from the wet condition. The daily soil water balance equation for the exposed and wetted soil fraction f_{ew} is:

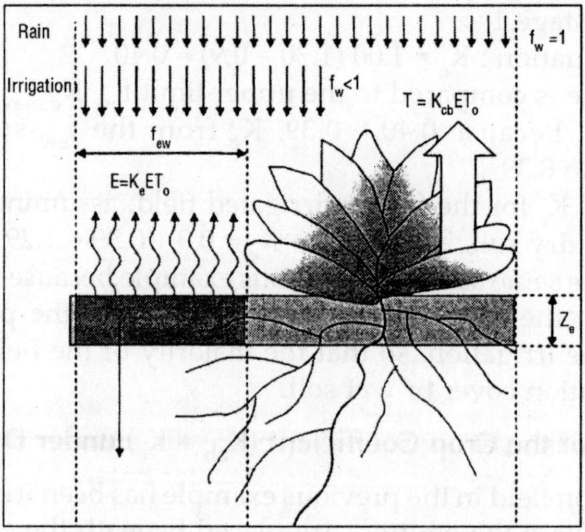

Fig. Water Balance of the Topsoil Layer

$$D_{ei} = D_{ei-1} - (P_i - RO_i) - \frac{I_i}{f_w} + \frac{E_i}{f_{ew}} + T_{ewi} + DP_{ei}$$

where, $D_{e,i-1}$ cumulative depth of evaporation following complete wetting from the exposed and wetted fraction of the topsoil at the end of day i – 1 [mm],

$D_{e,i}$ cumulative depth of evaporation (depletion) following complete wetting at the end of day i [mm],

P_i precipitation on day i [mm],

RO_i precipitation run off from the soil surface on day i [mm],

I_i irrigation depth on day i that infiltrates the soil [mm],

E_i evaporation on day i (i.e., $E_i = K_e ET_o$) [mm],

$T_{ew,i}$ depth of transpiration from the exposed and wetted fraction of the soil surface layer on day i [mm],

$DP_{e,i}$ deep percolation loss from the topsoil layer on day i if soil water content exceeds field capacity [mm], f_w fraction of soil surface wetted by irrigation [0.01 – 1],

f_{ew} exposed and wetted soil fraction [0.01 – 1].

Limits on De, i

By assuming that the topsoil is at field capacity following heavy rain or irrigation, the minimum value for the depletion $D_{e,i}$ is zero. As the soil surface dries, $D_{e,i}$ increases and in absence of any wetting event will steadily reach its maximum value TEW. At that moment no water is left for evaporation in the upper soil layer, K_r becomes zero, and the value for $D_{e,i}$ remains at TEW until the topsoil is wetted once again. The limits imposed on $D_{e,i}$ are consequently:

$$0 \leq D_{e,i} \leq TEW$$

Initial Depletion

To initiate the water balance for the evaporating layer, the user can assume that the topsoil is near field capacity following a heavy rain or irrigation, i.e., $D_{e,i-1} = 0$. Where a long period of time has elapsed since the last wetting, the user can assume that all evaporable water has been depleted from the evaporation layer at the beginning of calculations, i.e., $D_{e,i-1} = TEW = 1000\,(\theta_{FC} - 0.5\,\theta_{WP})\,Z_e$.

Precipitation and Runoff

P_i is equivalent to daily precipitation. Daily precipitation in amounts less than about 0.2 ET_o is normally entirely evaporated and can usually be ignored in the K_e and water balance calculations. The amount of rainfall lost by runoff depends on: the intensity of rainfall; the slope of land; the soil type, its hydraulic conditions and antecedent moisture content; and the land use and cover.

For general situations, RO_i can be assumed to be zero or can be accounted for by considering only a certain percentage of P_i. This is especially true for the water balance of the topsoil layer, since almost all precipitation events that would have intensities or depths large enough to cause runoff would probably replenish the water content of the topsoil layer to field capacity. Therefore, the impact of the runoff component can be ignored. Light precipitation events will generally have little or no runoff.

Irrigation

I_i is generally expressed as a depth of water that is equivalent to the mean infiltrated irrigation depth distributed over the entire field. Therefore, the value I_i/f_w is used to describe the actual concentration of the irrigation volume over the fraction of the soil that is wetted.

Evaporation

Evaporation beneath the vegetation canopy is assumed to be

included in K_{cb} and is therefore not explicitly quantified. The computed evaporation is fully concentrated in the exposed, wetted topsoil. The evaporation E_i is given by K_e ET_o. The E_i/f_{ew} provides for the actual concentration of the evaporation over the fraction of the soil that is both exposed and wetted.

Transpiration

Except for shallow rooted crops (i.e., where the depth of the maximum rooting zone is < 0.5 to 0.6 m), the amount of transpiration from the evaporating soil layer is small and can be ignored (i.e., T_{ew} = 0). In addition, for row crops, most of the water extracted by the roots may be extracted from beneath the vegetation canopy. Therefore, T_{ew} from the f_{ew} fraction of soil surface can be assumed to be zero in these cases.

Example. Estimation of crop evapotranspiration with the dual crop coefficient approach

- Estimate the crop evapotranspiration, ET_c, for ten successive days. It is assumed that:
  - The soil is a sandy loam soil, characterized by q_{FC} = 0.23 m^3 m^{-3} and q_{WP} = 0.10 m^3 m^{-3}.
  - The depth of the surface soil layer that is subject to drying by way of evaporation, Z_e, is 0.1 m.
  - During the period, the height of the vegetation h = 0.30 m, the average wind speed u_2 = 1.6 m s^{-1}, and RH_{min} = 35%.
  - The K_{cb} on day 1 is 0.30 and increases to 0.40 by day 10.
  - The exposed soil fraction, $(1 - f_c)$, decreases from 0.92 on day 1 to 0.86 on day 10.
  - All evaporable water has been depleted from the evaporation layer at the beginning of calculations ($D_{e, i-1}$ = TEW).
  - Irrigation occurs at the beginning of day 1 (I = 40 mm), and the fraction of soil surface wetted by irrigation, f_w = 0.8.
  - A rain of 6 mm occurred at the beginning of day 6.

From Table REW ≈ 8 mm.

- From Equation TEW = 1000 (0.23-0.5(0.10)) 0.1 = 18 mm.
- From Equation $K_{c\ max}$ = 1.2 + [0.04(1.6 - 2) – 0.004(35 – 45)] $(0.3/3)^{0.3}$ = 1.21.

All evaporable water has been depleted at the beginning of calculations, $D_{e,\ i-1}$ = TEW = 18 mm.

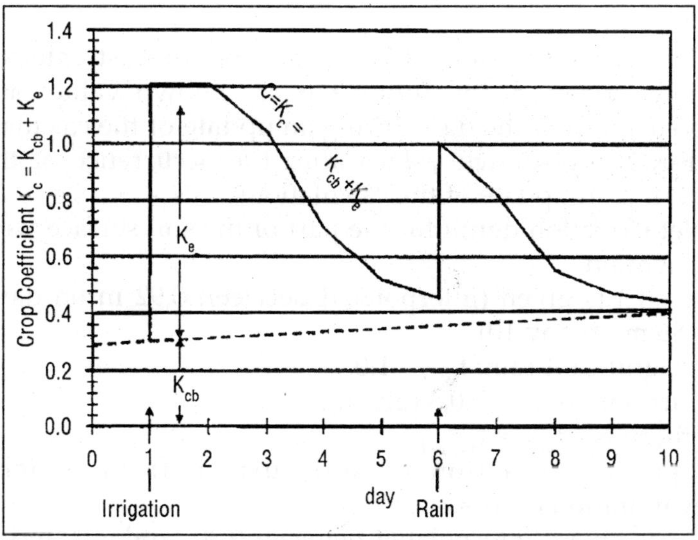

(1) Day	(2) ET_o	(3) P-RO	(4) I/f_w	(5) 1 - f_c	(6) f_w	(7) f_{ew}	(8) K_{cb}	(9) $D_{e,}$ start	(10) K_r	(11) K_e	(12) E/f_{ew}	(13) DP_e	(14) $D_{e,i}$ end	(15) E	(16) K_c	(17) ET_c
	mm/d	mm	mm					mm			mm	mm	mm	mm/d		mm/d
start	-	-	-	-	-	-	-	-	-	-	-	-	18	-	-	-
1	4.5	0	50	0.92	0.8	0.80	0.3[illegible]	[illegible]	1.00	0.91	5.1	32	5	4.1	1.21	5.5
2	5.0	0	0	0.91	0.8	0.80	0.31	5	1.00	0.90	5.6	0	11	4.5	1.21	6.1
3	3.9	0	0	0.91	0.8	0.80	0.32	11	0.70	0.62	3.0	0	14	2.8	1.04	4.0
4	4.2	0	0	0.90	0.8	0.80	0.33	14	0.40	0.35	1.8	0	16	1.5	0.70	2.9
5	4.8	0	0	0.89	0.8	0.80	0.34	1[illegible]	0.20	0.18	1.1	0	17	0.8	0.52	2.5
6	2.7	6	0	0.89	1	0.89	0.36	1[illegible]	0.75	0.64	2.0	0	13	1.7	1.00	2.7
7	5.8	0	0	0.88	1	0.88	0.37	13	0.53	0.45	3.0	0	16	2.6	0.82	4.7
8	5.1	0	0	0.87	1	0.87	0.38	16	0.20	0.17	1.0	0	17	0.9	0.55	2.8
9	4.7	0	0	0.87	1	0.87	0.39	17	0.09	0.08	0.4	0	18	0.4	0.47	2.2
10	5.2	0	0	0.86	1	0.86	0.40	18	0.05	0.04	0.2	0	18	0.2	0.44	2.3

- Day number.
- ET_o is given. Note that ET_o would be forecast values in real time irrigation scheduling but are known values after the occurrence of the day, during an update of the calculations.
- (P-RO) are known values after the occurrence of the day, during an update of the calculations.
- Net irrigation depth for the part of the soil surface wetted by irrigation.
- $(1 - f_c)$ is given (interpolated between 0.92 m on day 1 and 0.86 m on day 10).
- If significant rain: $f_{w,i} = 1.0$.
  If irrigation: $f_{w,i} = 0.8$ (given).
  otherwise: $f_{w,i} = f_{w,i-1}$.
- Equation. Fraction of soil surface from which most evaporation occurs.
- K_{cb} is given (interpolated between 0.30 on day 1 and 0.40 on day 10).
- $D_{e,i\,start}$ (depletion at start of day).

If precipitation and irrigation occur early in the day then the status of depletion from the soil surface layer (at the start of the day) should be updated:

$$= \text{Max}(D_{e,i-1} - I_{n,i}/f_{wi} - (P - RO)_i, \text{ or } 0).$$

where, $D_{e,i-1}$ is taken from column 14 of previous day.

If precipitation and irrigation occur late in the day, then column 6 should be set equal to $D_{e,i-1}$ (column 14 of previous day).

- If $D_{e,i} \le REW$ $K_r = 1$
  If $D_{e,i} > REW$ K_r = Equation.
- Equation where $K_e = K_r (K_{c\,max} - K_{cb}) \le f_{ew} K_{c\,max}$. (e.g., $K_e = \min(K_r (K_{c\,max} - K_{cb}), f_{ew} K_{c\,max})$.
- Evaporation from the wetted and exposed fraction of the soil surface = $(K_e\, ET_o)/f_{ew}$.
- Equation where $DP_e \ge 0$. (This is deep percolation from the evaporating layer).
- $D_{e,i}$ (depletion at end of day) is from Equation where $D_{e,i-1}$ is value in column 14 of previous day.
- Mean evaporation expressed as distributed over the entire field surface = $K_e\, ET_o$.
- $K_c = K_{cb} + K_e$.

The daily water balance calculation for the surface layer, even for shallow rooted crops, is not usually sensitive to T_{ew}, as T_{ew} is a minor part of the flux from the Z_e depth for the first 3-5 days following a wetting event. T_{ew} can, therefore, generally be ignored.

The effects of the reduction of the water content of the evaporating soil layer due to T_{ew} can be accounted for ulteriorly when it is assumed that $T_{ew} = 0$ by decreasing the value for Z_e, for example from 0.15 to 0.12 m or from 0.10 to 0.08 m.

Deep Percolation

Following heavy rain or irrigation, the soil water content in the topsoil (Z_e layer) might exceed field capacity. However, in this simple procedure it is assumed that the soil water content is at θ_{FC} nearly immediately following a complete wetting event, so that the depletion $D_{e,i}$ in Equation is zero. Following heavy rain or irrigation, downward drainage (percolation) of water from the topsoil layer is calculated as:

$$D_{ei} = (P_i - RO_i) + \frac{l_i}{f_w} - D_{ei-1} \geq 0$$

As long as the soil water content in the evaporation layer is below field capacity (i.e., $D_{e,i} > 0$), the soil will not drain and $DP_{e,i} = 0$.

Order of Calculation

In making calculations for the $K_{cb} + K_e$ procedure, for example when using a spreadsheet, the calculations should proceed in the following order: K_{cb}, h, $K_{c\,max}$, f_c, f_w, f_{ew}, K_r, K_e, E, DP_e, D_e, I, K_c, and ET_c.

CALCULATING ET_c

The calculation procedure lends itself to application by computer, either in the form of electronic spreadsheets or in the form of structured programming languages. The calculation procedure consists in determining:

A. Reference Evaporation, ET_o

Growth Stages

- Determine the locally adjusted lengths of the four growth stages:
  - Initial growth stage: L_{ini}.
  - Crop development stage: L_{dev}.
  - Mid-season stage: L_{mid}.
  - Late season stage: L_{late}.

C. Basal Crop Coefficient, K_{cb}

- Calculate basal crop coefficients for each day of the growing period:

- Select $K_{cb\ ini}$, $K_{cb\ mid}$ and $K_{cb\ end}$.
- Adjust $K_{cb\ mid}$ and $K_{cb\ end}$ to the local climatic conditions.
- Determine the daily K_{cb} values.

- *Initial growth stage:* $K_{cb} = K_{cb\ ini}$,
- *Crop development stage:* From $K_{cb\ ini}$ to $K_{cb\ mid}$,
- *Mid-season stage:* $K_{cb} = K_{cb\ mid}$,
- *Late season stage:* From $K_{cb\ mid}$ to $K_{cb\ end}$.

D. Evaporation Coefficient, K_e

- Calculate the maximum value of K_c, i.e., the upper limit $K_{c\ max}$, and Determine for each day of the growing period:
  - The fraction of soil covered by vegetation, f_c.
  - The fraction of soil surface wetted by irrigation or precipitation, f_w.
  - The fraction of soil surface from which most evaporation occurs, f_{ew}.
  - The cumulative depletion from the evaporating soil layer, D_e, determined by means of a daily soil water balance of the topsoil.
  - The corresponding evaporation reduction coefficient, K_r.
  - The soil evaporation coefficient, K_e.

E. Crop Evapotranspiration, ET_c

Calculate $ET_c = (K_{cb} + K_e)\ ET_o$.

Case Study of Dry Bean crop at Kimberly, Idaho, the United States (dual crop coefficient)

Results from applying the $K_{cb} + K_e$ procedure for a snap bean crop harvested as dry seed are shown. This example uses the same data set that was used in the case study. The measured ET_c data were measured using a precision lysimeter system at Kimberly, Idaho. Values for $K_{cb\ ini}$, $K_{cb\ mid}$, and $K_{cb\ end}$ were calculated in Example. The lengths of growth stages were 25, 25, 30, and 20 days. The K_{cb} values are plotted.

The value for $K_{c\ max}$ from Equation for the mid-season period averaged 1.24, based on u_2 = 2.2 m/s and RH_{min} = 30% for Kimberly. The soil at Kimberly was a silt loam texture.

Assuming that the depth of the evaporation soil layer, Z_e, was 0.1 m, values for TEW = 22 mm and REW = 9 mm, based on Equation and using soil data from table.

The occurrence and magnitudes of individual wetting events are shown in the Figure below. Nearly all wetting events were from irrigation. Because the irrigation was by furrow irrigation of alternate rows, the value for f_w was set equal to 0.5. Irrigation events occurred at about midday or during early afternoon. The agreement between the estimated values for daily $K_{cb} + K_e$ (thin continuous line) and actual 24-hour measurements (symbols) is relatively good. Measured and predicted $K_{cb} + K_e$ was higher following wetting by rainfall or irrigation, as expected. The two wet soil evaporation 'spikes' occurring during the late initial period and early development period (between days 160 and 180) were less than $K_{c\,max}$, because this evaporation was from wetting by furrow irrigation where $f_w = 0.5$.

The value for f_{ew} was constrained to f_w by Equation during these two events, because during this period, $f_w < 1 - f_c$. Therefore, less than all of the 'potential energy' was converted into evaporation due to the limitation on maximum evaporation per unit surface area that was imposed by Equation.

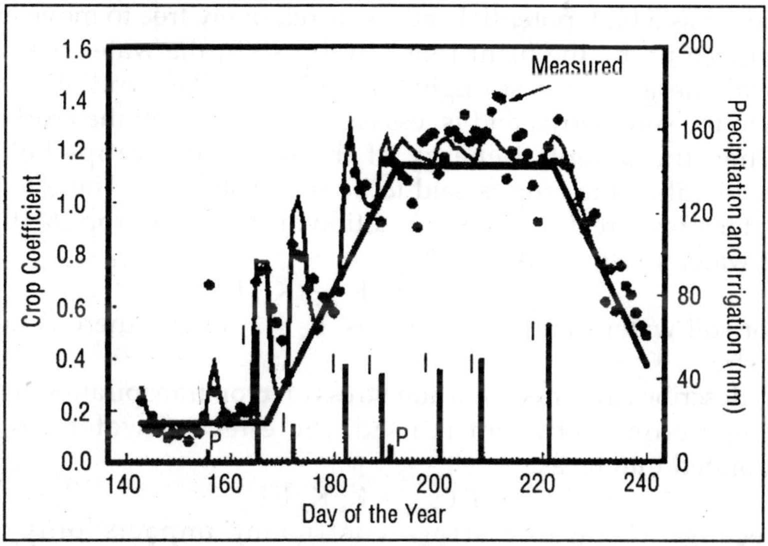

Measured (symbols) and predicted (thin line) daily coefficients ($K_{cb} + K_c$) and the basal crop curve (thick line) for a dry bean crop at Kimberly, Idaho. P in the Fig. denotes a precipitation event and I denotes an irrigation).

Chapter 7

Soil Water Under Stress Conditions

Forces acting on the soil water decrease its potential energy and make it less available for plant root extraction. When the soil is wet, the water has a high potential energy, is relatively free to move and is easily taken up by the plant roots. In dry soils, the water has a low potential energy and is strongly bound by capillary and absorptive forces to the soil matrix, and is less easily extracted by the crop.

When the potential energy of the soil water drops below a threshold value, the crop is said to be water stressed. The effects of soil water stress are described by multiplying the basal crop coefficient by the water stress coefficient, K_s:

$$ET_{c\ adj} = (K_s\ K_{cb} + K_e)\ ET_o$$

For soil water limiting conditions, $K_s < 1$. Where there is no soil water stress, $K_s = 1$.

K_s describes the effect of water stress on crop transpiration. Where the single crop coefficient is used, the effect of water stress is incorporated into K_c as:

$$ET_{c\ adj} = K_s\ K_c\ ET_o$$

Because the water stress coefficient impacts only crop transpiration, rather than evaporation from soil, the application of K_s using Equation is generally more valid than is application using Equation. However, in situations where evaporation from soil is not a large component of ET_c, use of Equation 81 will provide reasonable results.

SOIL WATER AVAILABILITY

Total Available Water (TAW)

Soil water availability refers to the capacity of a soil to retain water available to plants. After heavy rainfall or irrigation, the soil will drain until field capacity is reached. Field capacity is the amount of water that a well-drained soil should hold against gravitational forces, or the amount of water remaining when downward drainage has markedly decreased. In the absence of water supply, the water content in the root zone decreases as a result of water uptake by the crop.

As water uptake progresses, the remaining water is held to the soil particles with greater force, lowering its potential energy and making it more difficult for the plant to extract it. Eventually, a point is reached where the crop can no longer extract the remaining water. The water uptake becomes zero when wilting point is reached. Wilting point is the water content at which plants will permanently wilt.

As the water content above field capacity cannot be held against the forces of gravity and will drain and as the water content below wilting point cannot be extracted by plant roots, the total available water in the root zone is the difference between the water content at field capacity and wilting point:

$$TAW = 1000(\theta_{FC} - \theta_{WP})\, Z_r$$

where, TAW the total available soil water in the root zone [mm],

θ_{FC} the water content at field capacity [$m^3\ m^{-3}$],

θ_{WP} the water content at wilting point [$m^3\ m^{-3}$],

Z_r the rooting depth [m].

TAW is the amount of water that a crop can extract from its root zone, and its magnitude depends on the type of soil and the rooting depth. Typical ranges for field capacity and wilting point are listed in Table for various soil texture classes. Ranges of the maximum effective rooting depth for various crops are given in Table.

Readily Available Water (RAW)

Although water is theoretically available until wilting point, crop water uptake is reduced well before wilting point is reached. Where the soil is sufficiently wet, the soil supplies water fast enough to meet the atmospheric demand of the crop, and water uptake equals ET_c.

As the soil water content decreases, water becomes more strongly bound to the soil matrix and is more difficult to extract. When the soil water content drops below a threshold value, soil water can no longer be transported quickly enough towards the roots to respond to the transpiration demand and the crop begins to experience stress. The fraction of TAW that a crop can extract from the root zone without suffering water stress is the readily available soil water:

$$RAW = p\ TAW$$

where, RAW the readily available soil water in the root zone [mm],

p average fraction of Total Available Soil Water (TAW) that can be depleted from the root zone before moisture stress (reduction in

ET) occurs [0-1]. Values for p are listed in table. The factor p differs from one crop to another. The factor p normally varies from 0.30 for shallow rooted plants at high rates of ET_c (> 8 mm d^{-1}) to 0.70 for deep rooted plants at low rates of ET_c (< 3 mm d^{-1}). A value of 0.50 for p is commonly used for many crops.

The fraction p is a function of the evaporation power of the atmosphere. At low rates of ET_c, the p values listed in Table are higher than at high rates of ET_c. For hot dry weather conditions, where ET_c is high, p is 10-25% less than the values presented in Table, and the soil is still relatively wet when the stress starts to occur.

When the crop evapotranspiration is low, p will be up to 20% more than the listed values. Often, a constant value is used for p for a specific growing period, rather than varying the value each day. A numerical approximation for adjusting p for ET_c rate is $p = p + 0.04\,(5 - ET_c)$ where the adjusted p is limited to $0.1 \leq p \leq 0.8$ and ET_c is in mm/day. The influence of the numerical adjustment is shown in figure.

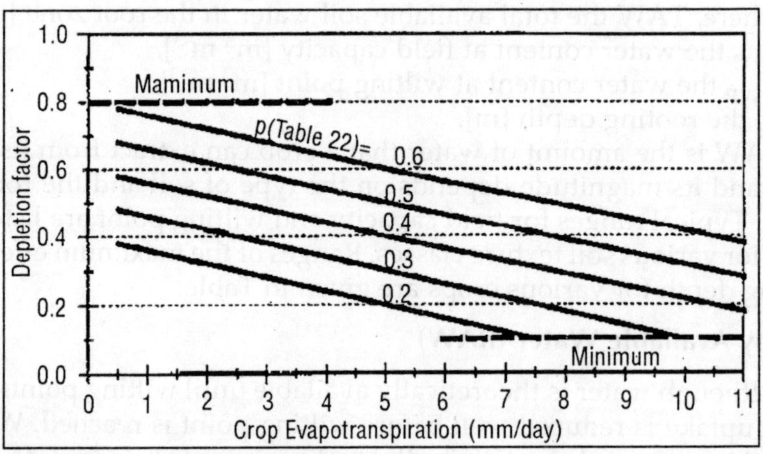

Fig. Depletion Factor for Different Levels of Crop Evapotranspiration.

To express the tolerance of crops to water stress as a function of the fraction (p) of TAW is not wholly correct. The rate of root water uptake is in fact influenced more directly by the potential energy level of the soil water (soil matric potential and the associated hydraulic conductivity) than by water content.

As a certain soil matric potential corresponds in different soil types with different soil water contents, the value for p is also a function of the soil type. Generally, it can be stated that for fine textured soils (clay) the p values listed in Table can be reduced by 5-10%, while for more coarse textured soils (sand), they can be increased by 5-10%. RAW is similar to the term Management Allowed Depletion (MAD) introduced by Merriam However, values for MAD are influenced by management and economic factors in addition to the physical factors influencing p. Generally, MAD < RAW where there is risk aversion or uncertainty, and MAD > RAW where plant moisture stress is an intentional part of soil water management.

Determination of Readily Available Soil Water for Various Crops and Soil Types.

Estimate RAW for a full-grown onion, tomato and maize crop. Assume that the crops are cultivated on loamy sand, silt and silty clay soils.						
From Table	Onion		$Z_r \approx 0.4$ m		$p = 0.30$	
	Tomato		$Z_r \approx 0.8$ m		$p = 0.40$	
	Maize		$Z_r \approx 1.2$ m		$p = 0.55$	
From Table	Loamy sand		$\theta_{FC} \approx 0.15$ m^3 m^{-3}		$\theta_{WP} \approx 0.06$ m^3 m^{-3}	
	$1000\ (\theta_{FC} - \theta_{WP}) = 90$ mm(water)/m(soil depth)					
	Silt		$\theta_{FC} \approx 0.32$ m^3 m^{-3}		$\theta_{WP} \approx 0.15$ m^3 m^{-3}	
	$1000\ (\theta_{FC} - \theta_{WP}) = 170$ mm(water)/m(soil depth)					
	Silty clay		$\theta_{FC} \approx 0.35$ m^3 m^{-3}		$\theta_{WP} \approx 0.23$ m^3 m^{-3}	
	$1000\ (\theta_{FC} - \theta_{WP}) = 120$ mm(water)/m(soil depth)					
	Loamy sand		**Silt**		**Silty clay**	
	TAW (Eq. 82) mm	**RAW (Eq. 83) mm**	**TAW (Eq. 82) mm**	**RAW (Eq. 83) mm**	**TAW (Eq. 82) mm**	**RAW (Eq. 83) mm**
Onion	36	11	68	20	48	14
Tomato	72	29	136	54	96	38
Maize	108	59	204	112	144	79

WATER STRESS COEFFICIENT (K_s)

The effects of soil water stress on crop ET are described by reducing the value for the crop coefficient. This is accomplished by multiplying the crop coefficient by the water stress coefficient, K_s.

Water content in me root zone can also be expressed by root zone depletion, D_r, i.e., water shortage relative to field capacity. At field capacity, the root zone depletion is zero ($D_r = 0$). When soil water is extracted by evapotranspiration, the depletion increases and stress will be induced when D_r becomes equal to RAW.

After the root zone depletion exceeds RAW (the water content drops below the threshold θ_t), the root zone depletion is high enough to limit evapotranspiration to less than potential values and the crop evapotranspiration begins to decrease in proportion to the amount of water remaining in the root zone.

Effect of Water Stress on Crop Evapotranspiration

Estimate the effect of water stress on the evapotranspiration of a full grown tomato crop ($Z_r = 0.8$ m and $p = 0.40$) cultivated on a silty soil (θ_{FC} = 0.32 and θ_{WP} = 0.12 $m^3 m^{-3}$) for the coming 10 days when the initial root zone depletion is 55 mm and neither rain nor irrigations are either forecasted or planned. The expected ET_o for the coming decade is 5 mm/day and K_c = 1.2.

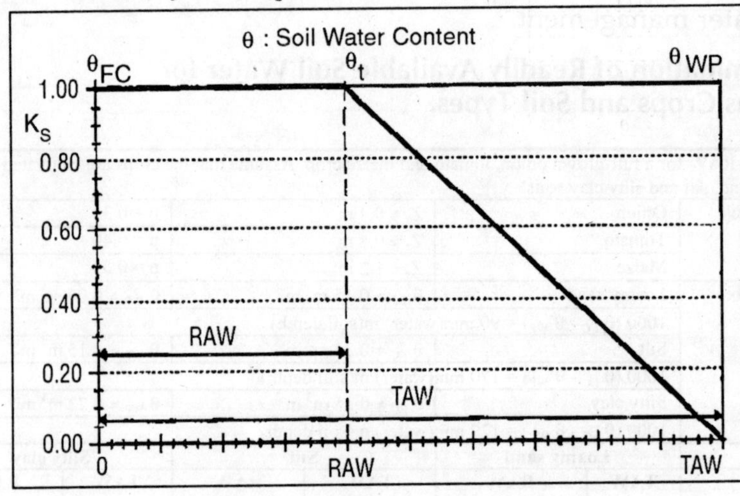

Fig. Water Stress Coefficient, K_s

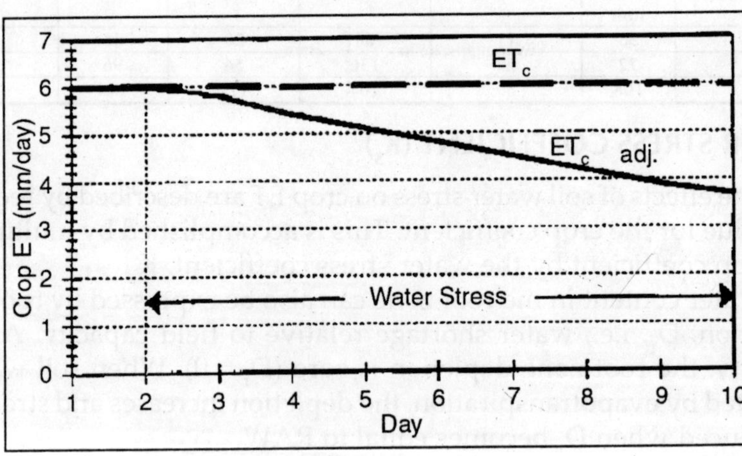

For D_r > RAW, K_s is given by:

$$K_s = \frac{TAW - D_r}{TAW - RAW} = \frac{TAW - D_r}{(1-p)TAW}$$

where, K_s is a dimensionless transpiration reduction factor dependent on available soil water [0 - 1],

D_r root zone depletion [mm],

TAW total available soil water in the root zone [mm],

p fraction of TAW that a crop can extract from the root zone without suffering water stress [-].

From Equation						TAW = 1000 (0.32-0.12) 0.8 = 160 mm	
From Equation						RAW = 0.40 (160) = 64 mm	
(1)	(2)	(3)	(4)	(5)	(6)	(7)	(8)
Day	ET_o	K_c	ET_c	$D_{r,i\ start}$	K_s	$ET_{c\ adj}$	$D_{r,i\ end}$
	mm/day		mm/day	mm		mm/day	mm
start	-	-	-	-	-	-	55.0
1	5.0	1.2	6.0	55.0	1.00	6.0	61.0
2	5.0	1.2	6.0	61.0	1.00	6.0	67.0
3	5.0	1.2	6.0	67.0	0.97	5.8	72.8
4	5.0	1.2	6.0	72.8	0.91	5.4	78.3
5	5.0	1.2	6.0	78.3	0.85	5.1	83.4
6	5.0	1.2	6.0	83.4	0.80	4.8	88.2
7	5.0	1.2	6.0	88.2	0.75	4.5	92.6
8	5.0	1.2	6.0	92.6	0.70	4.2	96.9
9	5.0	1.2	6.0	96.9	0.66	3.9	100.8
10	5.0	1.2	6.0	100.8	0.62	3.7	104.5
(1)	Day number.						
(2)	Reference crop evapotranspiration.						
(3)	Crop coefficient.						
(4)	Equation, crop ET if no water stress.						
(5)	Root zone depletion at the beginning of the day (column 8 of previous day).						
(6)	Equation where $K_s = 1$ if $D_{r,i} < RAW$.						
(7)	Equation, crop ET under soil water stress conditions.						
(8)	Depletion at end of day.						

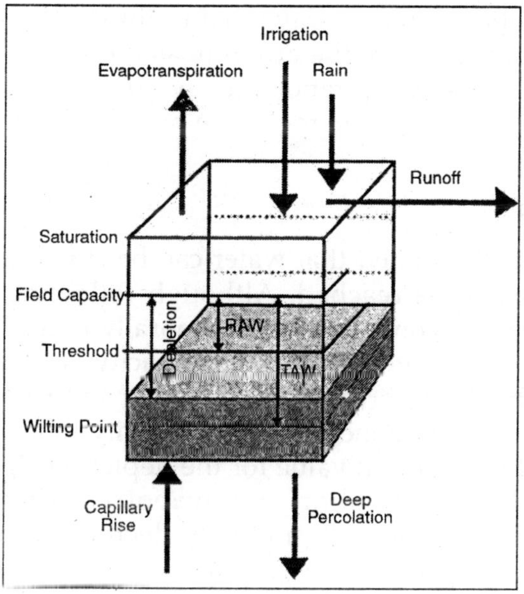

Fig. Water Balance of the Root Zone

After the computation of K_s, the adjusted evapotranspiration $ET_{c\ adj}$ is computed by means of Equation, depending on the coefficients used to describe crop evapotranspiration. When the root zone depletion is smaller than RAW, $K_s = 1$.

Soil Water Balance

The estimation of K_s requires a daily water balance computation for the root zone. Schematically, the root zone can be presented by means of a container in which the water content may fluctuate. To express the water content as root zone depletion is useful. It makes the adding and subtracting of losses and gains straightforward as the various parameters of the soil water budget are usually expressed in terms of water depth.

Rainfall, irrigation and capillary rise of groundwater towards the root zone add water to the root zone and decrease the root zone depletion. Soil evaporation, crop transpiration and percolation losses remove water from the root zone and increase the depletion. The daily water balance, expressed in terms of depletion at the end of the day is:

$$D_{r,i} = D_{r,i-1} - (P - RO)_i - I_i - CR_i + ET_{c,i} + DP_i$$

where, $D_{r,i}$ root zone depletion at the end of day i [mm],

$D_{r,i-1}$ water content in the root zone at the end of the previous day, i-1 [mm],

P_i precipitation on day i [mm],

RO_i runoff from the soil surface on day i [mm],

I_i net irrigation depth on day i that infiltrates the soil [mm],

CR_i capillary rise from the groundwater Table on day i [mm],

$ET_{c,i}$ crop evapotranspiration on day i [mm],

DP_i water loss out of the root zone by deep percolation on day i [mm].

Limits on Dr, i

In figure it is assumed that water can be stored in the root zone until field capacity is reached. Although following heavy rain or irrigation the water content might temporally exceed field capacity, the total amount of water above field capacity is assumed to be lost the same day by deep percolation, following any ET for that day. By assuming that the root zone is at field capacity following heavy rain or irrigation, the minimum value for the depletion $D_{r,i}$ is zero.

As a result of percolation and evapotranspiration, the water content in the root zone will gradually decrease and the root zone depletion will increase.

In the absence of any wetting event, the water content will steadily reach its minimum value θ_{WP}. At that moment no water is left for evapotranspiration in the root zone, K_s becomes zero, and the root zone depletion has reached its maximum value TAW. The limits imposed on $D_{r,i}$ are consequently:

$$0 \leq D_{r,i} \leq TAW$$

Initial Depletion

To initiate the water balance for the root zone, the initial depletion $D_{r,\,i-1}$ should be estimated. The initial depletion can be derived from measured soil water content by:

$$D_{r,\,i-1} = 1000(\theta_{FC} - \theta_{i-1})Z_r$$

where θ_{i-1} is the average soil water content for the effective root zone. Following heavy rain or irrigation, the user can assume that the root zone is near field capacity, i.e., $D_{r,\,i-1} \approx 0$.

Precipitation (P), Runoff (RO) and Irrigation (I)

P_i is equivalent to daily precipitation. Daily precipitation in amounts less than about 0.2 ET_o is normally entirely evaporated and can usually be ignored in the water balance calculations especially when the single crop coefficient approach is being used. Ii is equivalent to the mean infiltrated irrigation depth expressed for the entire field surface. Runoff from the surface during precipitation can be predicted using standard procedures from hydrological texts.

Capillary Rise (CR)

The amount of water transported upwards by capillary rise from the water Table to the root zone depends on the soil type, the depth of the water Table and the wetness of the root zone. CR can normally be assumed to be zero when the water Table is more than about 1 m below the bottom of the root zone.

Some information on CR was presented in FAO Irrigation and Drainage Paper. CR will be a topic in a future FAO publication.

Evapotranspiration (ET_c)

Where the soil water depletion is smaller than RAW, the crop evapotranspiration equals $ET_c = K_c\, ET_o$. As soon as $D_{r,\,i}$ exceeds RAW, the crop evapotranspiration is reduced and ET_c can be computed from Equation.

Deep Percolation (DP)

Following heavy rain or irrigation, the soil water content in the root zone might exceed field capacity. In this simple procedure it is assumed that the soil water content is at θ_{FC} within the same day of the wetting event, so that the depletion $D_{r,\,i}$ in Equation becomes zero. Therefore, following heavy rain or irrigation

$$DP_i = (P_i - RO_i) + I_i - ET_{c,\,i} - D_{r,\,i-1} \geq 0$$

As long as the soil water content in the root zone is below field capacity (i.e., $D_{r,\,i} > 0$), the soil will not drain and $DP_i = 0$.

The DP_i term in Equations 85 and 88 is not to be confused with the $DP_{e,i}$ term used in Equations for the evaporation layer. Both terms can be calculated at the same time, but are independent of one another.

Forecasting or Allocating Irrigations

Irrigation is required when rainfall is insufficient to compensate for the water lost by evapotranspiration. The primary objective of irrigation is to apply water at the right period and in the right amount. By calculating the soil water balance of the root zone on a daily basis, the timing and the depth of future irrigations can be planned.

To avoid crop water stress, irrigations should be applied before or at the moment when the readily available soil water is depleted ($D_{r,i} \leq RAW$).

To avoid deep percolation losses that may leach relevant nutrients out of the root zone, the net irrigation depth should be smaller than or equal to the root zone depletion ($I_i \leq D_{r,i}$).

Example. Illustrates the application of a water balance of the root zone to predict irrigation dates to avoid water stress. The example utilizes various calculations for K_e from Example. A complete "spreadsheet" that includes all necessary calculations for predicting both irrigation schedules and to predict $K_c = K_{cb} + K_e$ for daily timesteps.

Irrigation Scheduling to Avoid Crop Water Stress

- Plan the irrigation applications for Example. It is assumed that:
  - Irrigations are to be applied when RAW is depleted.
  - The depletion factor (p) is 0.6.
  - All irrigations and precipitations occur early in the day.
  - The depth of the root zone (Z_r) on day 1 is 0.3 m and increases to 0.35 m by day 10.
  - The root zone depletion at the beginning of day 1 ($D_{r,i-1}$) is RAW.

From Equation $TAW = 1000\,(0.23 - 0.10)\,Z_{r,i} = 130\,Z_{r,i}$ [mm]

From Equation $RAW = 0.6\,TAW = 78\,Z_{r,i}$ [mm]

On day 1, when $Z_r = 0.3$ m: $D_{r,i-1} = RAW = 78\,(0.3) = 23$ mm

Effects of Soil Salinity

Salts in the soil water solution can reduce evapotranspiration by making soil water less "available" for plant root extraction. Salts have an affinity for water and hence additional force is required for the crop to extract water from a saline soil. The presence of salts in the soil water

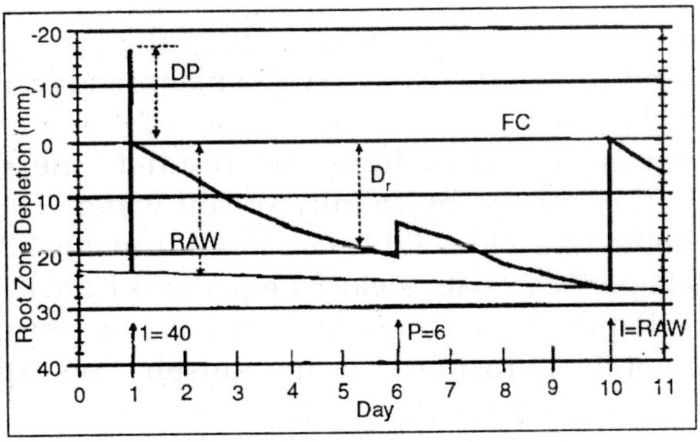

(1) Day	(2) ET_o	(3) Z_r	(4) RAW	(5) $D_{r,i\ start}$	(6) P-RO	(7) I	(8) K_s	(9) K_{cb}	(10) K_e	(11) K_c	(12) ET_c	(13) DP	(14) $D_{r,i\ end}$
	mm/d	m	mm	mm	mm	mm					mm	mm	mm
start	-	-	-	-	-	-	-	-	-	-	-	-	23
1	4.5	0.30	23	0	0	40	1	0.30	0.91	1.21	5.5	17	5
2	5.0	0.31	24	5	0	0	1	0.31	0.90	1.21	6.1	0	12
3	3.9	0.31	24	12	0	0	1	0.32	0.72	1.04	4.0	0	16
4	4.2	0.32	25	16	0	0	1	0.33	0.37	0.70	2.9	0	18
5	4.8	0.32	25	18	0	0	1	0.34	0.18	0.52	2.5	0	21
6	2.7	0.33	26	15	6	0	1	0.36	0.64	1.00	2.7	0	18
7	5.8	0.33	26	18	0	0	1	0.37	0.45	0.82	4.7	0	22
8	5.1	0.34	26	22	0	0	1	0.38	0.17	0.55	2.8	0	25
9	4.7	0.34	27	25	0	0	1	0.39	0.08	0.47	2.2	0	27
10	5.2	0.35	27	0	0	27	1	0.40	0.81	1.21	6.3	0	6

solution reduces the total potential energy of the soil water solution. In addition, some salts cause toxic effects in plants and can reduce plant metabolism and growth.

A function is presented here that predicts the reduction in evapotranspiration caused by salinity of soil water. The function is derived by combining yield-salinity equations from the FAO Irrigation and Drainage Paper N°29 with yield-ET equations from FAO Irrigation and Drainage Paper N°33. The resulting equation provides a first approximation of the reduction in evapotranspiration expected under various salinity conditions.

There is evidence that crop yield and transpiration are not as sensitive to low osmotic potential as they are to low matric potential. Under saline conditions, many plants are able to partially compensate for low osmotic potential of the soil water by building up higher internal solute contents.

This is done by absorbing ions from the soil solution and by synthesizing organic osmolytes. Both of these reactions reduce the impact of osmotic potential on water availability. However, synthesis of organic osmolytes does require expenditure of metabolic energy. Therefore plant growth is often reduced under saline conditions. The reduced plant growth impacts transpiration by reducing ground cover and is sometimes additionally due to partial stomatal closure.

Other impacts of salts in the soil include direct sodium and chloride toxicities and induced nutrient deficiencies. These deficiencies reduce plant growth by reducing the rate of leaf elongation, the enlargement, and the division of cells in leaves. The modality depends on the method of irrigation. With sprinkler irrigation, adsorption of sodium and chloride through the leaf can result in toxic conditions for all crop species. With surface or trickle irrigation, direct toxic conditions generally occur only in vine and tree crops; however, high levels of sodium can induce calcium deficiencies for all crop species.

Since salt concentration changes as the soil water content changes, soil salinity is normally measured and expressed on the basis of the electrical conductivity of the saturation extract of the soil (EC_e). The EC_e is defined as the electrical conductivity of the soil water solution after the addition of a sufficient quantity of distilled water to bring the soil water content to saturation. EC_e is typically expressed in deciSiemens per meter (dS m^{-1}).

Under optimum management conditions, crop yields remain at potential levels until a specific, threshold electrical conductivity of the saturation soil water extract ($EC_{e\ threshold}$) is reached. If the average EC_e of the root zone increases above this critical threshold value, the yield is presumed to begin to decrease linearly in proportion to the increase in salinity. The rate of decrease in yield with increase in salinity is usually expressed as a slope, b, having units of% reduction in yield per dS/m increase in EC_e.

All plants do not respond to salinity in a similar manner; some crops can produce acceptable yields at much higher soil salinity levels than others. This is because some crops are better able to make the needed osmotic adjustments that enable them to extract more water from a saline soil, or they may be more tolerant of some of the toxic effects of salinity.

As can be observed there is an 8 to 10-fold range in salt tolerance of agricultural crops. The effect of soil salinity on yield and crop evapotranspiration is hence crop specific.

The $EC_{e\ threshold}$ and b parameters were determined primarily in research experiments using nearly steady-state irrigation where soil water contents were maintained at levels close to field capacity. However, under most types of irrigation scheduling for sprinkler and surface irrigation, the soil water content is typically depleted to well below field capacity, so that the EC of the soil water solution, EC_{SW}, increases prior to irrigation, even though the EC of the saturation extract does not change.

The increased salt concentration in the soil water solution reduces the osmotic potential of the soil water solution (it becomes more negative), so that the plant must expend more metabolic energy and may exert more mechanical force to absorb water. In addition, metabolic and toxic effects of salts on plants may become more pronounced as the soil dries and concentrations increase.

However, the variation in soil water content during an irrigation interval has not been found to strongly influence crop evapotranspiration. This is because of the rise of soil water content to levels that are above that experienced under steady state irrigation early in a long irrigation interval. There is a similar, counteractive decrease in soil water content later in a long irrigation interval.

In addition, the distribution of salts in the root zone under low frequency irrigation can reduce salinity impacts during the first portion of the irrigation interval. Also, under high frequency irrigation of the soil surface, soil evaporation losses are higher. Consequently, given the same application depth, the leaching fraction is reduced. For these reasons, the length of irrigation interval and the change in EC of soil water during the interval have usually not been found to be factors in the reduction of ET, given that the same depths of water are infiltrated into the root zone over time.

In some cases, increased evaporation under high frequency irrigation of the soil surface can counteract reductions in K_c caused by high EC_e of the root zone. Under these conditions, the total K_c and ET_c are not very different from the non-saline, standard conditions under less frequent irrigation, even though crop yields and crop transpiration are reduced. Because of this, under saline conditions, the K_s reducing factor should only be applied with the dual K_c approach.

In review articles on impacts of salinity on crop production, Letey *et al.* and Shalhevet concluded that effects of soil salinity and water stress are generally additive in their impacts on crop evapotranspiration. Therefore, the same yield-ET functions may hold for both water shortage induced stress and for salinity induced stress.

Yield-Salinity Relationship

A widely practiced approach for predicting the reduction in crop yield due to salinity has been described in the FAO Irrigation and Drainage Paper N°29. The approach presumes that, under optimum management conditions, crop yields remain at potential levels until a specific, threshold electrical conductivity of the soil water solution is reached.

When salinity increases beyond this threshold, crop yields are presumed to decrease linearly in proportion to the increase in salinity. The soil water salinity is expressed as the electrical conductivity of the saturation extract, EC_e. In equation form, the procedure followed in FAO Irrigation and Drainage Paper N°29 is:

$$\frac{Y_a}{Y_m} = 1 - (EC_e - EC_{etreshold})\frac{b}{100}$$

for conditions where $EC_e > EC_{e\ threshold}$ where:

Y_a actual crop yield

Y_m maximum expected crop yield when $EC_e < EC_{e\ threshold}$

EC_e mean electrical conductivity of the saturation extract for the root zone [dS m^{-1}]

$EC_{e\ threshold}$ electrical conductivity of the saturation extract at the threshold of EC_e when crop yield first reduces below Y_m [dS m^{-1}]

b reduction in yield per increase in EC_e [%/(dS m^{-1})]

Values for $EC_{e\ threshold}$ and b have been provided in the FAO Irrigation and Drainage Paper N°29 and 48 and are listed in Table for many agricultural crops.

Salinity-yield data from the FAO Irrigation and Drainage papers were mostly from studies where soil water content was held at about-3 m potential (–30 kPa) or higher at the 0.3 to 0.6 m depth, depending on the crop. However, these papers indicate that for most crops, the data are transferable to typical field situations where the readily available soil water (RAW) is depleted between irrigations.

Yield-Moisture Stress Relationship

A simple, linear crop-water production function was introduced in the FAO Irrigation and Drainage Paper N°33 to predict the reduction in crop yield when crop stress was caused by a shortage of soil water:

$$\left(1-\frac{Y_a}{Y_m}\right)=K_y\left(1-\frac{ET_{cadj}}{ET_c}\right)$$

where: K_y a yield response factor [-]

$ET_{c\ adj}$ adjusted (actual) crop evapotranspiration [mm d^{-1}]

ET_c crop evapotranspiration for standard conditions (no water stress) [mm d^{-1}]

K_y is a factor that describes the reduction in relative yield according to the reduction in ET_c caused by soil water shortage. In FAO N°33, K_y values are crop specific and may vary over the growing season. In general, the decrease in yield due to water deficit during the vegetative and ripening period is relatively small, while during the flowering and yield formation periods it will be large.

Values for K_y for individual growth periods and for the complete growing season have been included in the FAO Irrigation and Drainage Paper N°33.

COMBINED SALINITY-ET REDUCTION RELATIONSHIP

No Water Stress (D_r < RAW)

When salinity stress occurs without water stress, Equations can be combined and solved for an equivalent K_s, where $K_s = ET_{c\ adj}/ET_c$:

$$K_s = 1-\frac{b}{K_y 100}(EC_e - EC_{ethreshold})$$

for conditions when $EC_e > EC_{e\ threshold}$ and soil water depletion is less than the readily available soil water depth (D_r < RAW). D_r and RAW are defined in the previous section.

With Water Stress (D_r > RAW)

When soil water stress occurs in addition to salinity stress.

$$K_s = \left(1-\frac{b}{K_y 100}(EC_e - EC_{etreshold})\right)\left(\frac{TAW - D_r}{TAW - RAW}\right)$$

For conditions when $EC_e > EC_{e\ threshold}$ and D_r > RAW. Figure shows the impact of salinity reduction on K_s as salinity increases. Note that the approach presumes that RAW (and p) do not change with increasing salinity. This may or may not be a good assumption for some crops.

Because the impact of salinity on plant growth and yield and on crop evapotranspiration is a time-integrated process, generally only the seasonal value for K_y is used to predict the reduction in evapotranspiration. There are K_y values in FAO Irrigation and Drainage paper N°33 for only about 23 crops.

The seasonal values for K_y from paper N°33 are summarized. For many crops, the seasonal K_y is nearly 1. For crops where K_y is unknown, the user may use $K_y = 1$ in Equations may select the K_y for a crop type that has similar behaviour.

Equations are suggested as only approximate estimates of salinity impacts on ET, and represent general effects of salinity on evapotranspiration as occurring over an extended period of time (as measured in weeks or months). These equations are not expected to be accurate for predicting ET_c for specific days. Nor do they include other complicating effects such as specific ion toxicity.

The equations presented may not be valid at high salinity, where the linear relationships between EC_e, crop yield and K_s may not hold. The use of Equations should usually be restricted to $EC_e < EC_{threshold} + 50/b$. In addition, the equations predict $Y_a = 0$ before $K_s = 0$ when $K_y > 1$ and vice versa. As indicated earlier, reduction in ET_c in the presence of soil salinity is often partially caused by reduced plant size and fraction of ground cover. These effects are largely included in the coefficient values in Table. Therefore, where plant growth is affected by salinity and Equations are applied, no other reductions in K_c are required, for example using LAI or fraction of ground cover.

FAO Irrigation and Drainage Paper N° 29. Water quality for agriculture; Rhoades, Kandiah and Mashali, 1992. FAO Irrigation and Drainage Paper N° 48. The use of saline waters for crop productions.

Application

Under steady state conditions, the value for EC_e can be predicted as a function of EC of the irrigation water (EC_{iw}) and the leaching fraction, using a standard leaching formula. For example, the FAO leaching formula $LR = EC_{iw}/(5\ EC_e - EC_{iw}$ predicts the leaching requirement when approximately a 40-30-20-10 per cent water extraction pattern occurs from the upper to lower quarters of the root zone prior to irrigation. EC_{iw} is the electrical conductivity of the irrigation water. From this equation, EC_e is estimated as:

$$EC_e = \frac{1+LF}{LF}\frac{ECiw}{5}$$

Table. Seasonal Yield Response Functions from FAO Irrigation and Drainage Paper

Crop	K_y
Alfalfa	1.1
Banana	1.2-1.35
Beans	1.15
Cabbage	0.95
Citrus	1.1-1.3
Cotton	0.85
Grape	0.85
Groundnet	0.70
Maize	1.25
Onion	1.1
Peas	1,15
Pepper	1.1
Potato	1.1
Safflower	0.8
Sorghum	0.9
Soybean	0.85
Spring Wheat	1.15
Sugarbeet	1.0
Sugarcane	1.2
Sunflower	0.95
Tomato	1.05
Watermelon	1.1
Winter wheat	1.05

Where LF, the actual leaching fraction, is used in place of LR, the leaching requirement.

Equation predicts $EC_e = 1.5\ EC_{iw}$ under conditions where a 15-20 per cent leaching fraction is employed. Other leaching fraction equations can be used in place of the FAO equation to fit local characteristics. Equation is only true if the irrigation water quality and the leaching fraction are constant over the growing season.

Time is required to attain a salt equilibrium in the soil. If there are important winter rains of high quality water and often excellent leaching, the salt balance in the soil will be quite different at the beginning of the season and with a lower average EC_e of the root zone than would be predicted from Equation. An appropriate local calibration of Equation is desirable under these particular conditions.

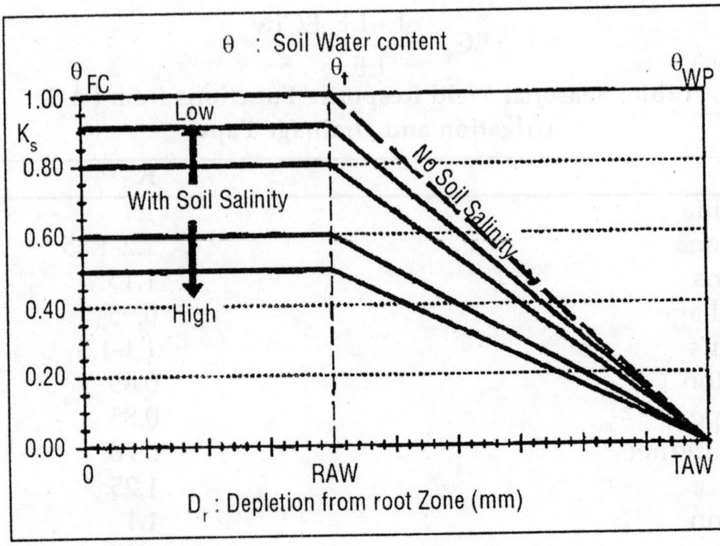

Fig. The Effect of Soil Salinity on the Water Stress Coefficient K_s

Effect of Soil Salinity on Crop Evapotranspiration

A field of beans is cultivated on a silt loam soil and is irrigated during the midseason period using water having salinity EC_{iw} = 1 dS m^{-1}. A 15 per cent leaching fraction is employed. The $EC_{e,\ threshold}$ and slope from Table are 1.0 dS m^{-1} and 19%/(dS m^{-1}) respectively. The seasonal K_y from FAO Irrigation and Drainage Paper for beans is K_y = 1.15. Compare the effect on crop evapotranspiration for various levels of soil water depletion in the root zone under saline and nonsaline conditions. The TAW and RAW for the bean crop are 110 and 44 mm (for p = 0.4).

Since the leaching fraction is 0.15, EC_e is estimated from Equation 93 as $EC_e = 1.5\ EC_w = 1.5\ (1) = 1.5$ dS m^{-1}. The K_s in the presence of salinity stress and absence of moisture stress is:

$$K_s = \left(1 - \frac{b}{K_y 100}(EC_e - EC_{etreshold})\right) = \left(1 - \frac{19}{1.15(100)}(1.5 - 1.0)\right) = 0.92$$

The K_s in the presence of moisture stress, but in the absence of salinity stress is:

$$K_s = \left(\frac{TAW - D_r}{TAW - RAW}\right) = \left(\frac{110 - D_r}{100 - 44}\right) = \left(\frac{110 - D_r}{66}\right)$$

The K_s in the presence of both moisture stress and salinity stress is:

$$K_s = \left(1 - \frac{b}{K_y 100}(EC_e - EC_{etreshold})\right)\left(\frac{TAW - D_r}{TAW - RAW}\right) = 0.92\left(\frac{110 - D_r}{66}\right)$$

The effect on crop evapotranspiration for various soil water depletions in the root zone (D_r) are:

D_r (mm)	K_s no soil salinity		K_s with soil salinity (EC_e = 1.5 dS m^{-1}) (Eq. 92)		Additional reduction in potential ET_c due to salinity
0	1.00	no reduction in ET_c	0.92	8% reduction in ET_c	8%
35	1.00	no reduction in ET_c	0.92	8% reduction in ET_c	8%
40	1.00	no reduction in ET_c	0.92	8% reduction	8%
44	1.00	no reduction in ET_c	0.92	8% reduction	8%
50	0.91	9% reduction	0.83	17% reduction	8%
60	0.76	24% reduction	0.69	31 % reduction	7%
70	0.61	39% reduction	0.56	44% reduction	5%
80	0.45	55% reduction	0.42	58% reduction	3%
90	0.30	70% reduction	0.28	72% reduction	2%
100	0.15	85% reduction	0.14	86% reduction	1%
110	0.00	ET_c = 0	0.00	ET_c = 0	--

Chapter 8

Non-Pristine Vegetation

Non-typical refers to types or arrangements of agricultural crops that are not listed. Non-pristine vegetation is defined, in the usage here, as vegetation having less than perfect growing conditions or stand characteristics (i.e., relatively poorer conditions of density, height, leaf area, fertility, or vitality) as compared to 'pristine' conditions.

The approach whereby a crop is characterized by a crop coefficient, K_c, and the crop evapotranspiration is given by the product of K_c and the reference evapotranspiration ET_o, provides a simple and convenient way of also characterizing the evapotranspiration from natural vegetation and for non-typical cultivation practices. This chapter presents procedures for estimating K_c values for natural vegetation and for agricultural vegetation for which K_c values are not available.

CALCULATION APPROACH

The first step in the K_cET_o approach is the estimation of lengths of growth stages. This also applies to natural and other vegetation. The next step is the development of crop coefficient curves that represent the ratios of ET_c to ET_o during the various growth stages of the vegetation.

Initial Growth Stage

The procedure to estimate crop coefficients for the initial growth stage for natural, non-typical and non-pristine vegetation is identical (single crop coefficient $K_{c\,ini}$) or (dual crop coefficient, $K_{cb\,ini} + K_e$). The crop coefficient in this stage is primarily determined by the frequency with which the soil is wetted.

Mid and Late Season Stages

The K_c during the mid-season period ($K_{c\,mid}$ and $K_{cb\,mid}$) and to a lesser extent the K_c during the late season period differ from that described in previous chapters. As the ground cover for natural and non-pristine vegetation is often reduced, the K_c is affected to a large extent by the frequency of precipitation and/or irrigation and by the amount of leaf area and ground cover.

Dual Crop Coefficient Approach

The determination of K_c for natural, non-typical or non-pristine vegetation should ordinarily follow the approach whereby separate transpiration (K_{cb}) and evaporation (K_e) coefficients are used. The effects of evaporation from the soil surface can be directly estimated as such.

Two procedures that can be used to adjust the basal crop coefficient ($K_{cb\ mid\ adj}$) for sparse vegetation are presented in this section. In these approaches, $K_{cb\ mid\ adj}$ is estimated either from LAI or from effective ground cover. After the determination of $K_{cb\ mid\ adj}$, the soil evaporation coefficient, K_e, should be determined to obtain the crop coefficient for the mid-season stage: $K_{c\ mid\ adj} = K_{cb\ mid\ adj} + K_e$. Procedures for calculating K_e.

Even where the estimated $K_{cb\ mid\ adj}$ is small, the total $K_{c\ adj}$ (= $K_{cb\ adj} + K_e$) following precipitation may sometimes be as high or higher than the K_c for pristine vegetation due to surface evaporation from among sparse vegetation.

Single Crop Coefficient Approach

When the single crop coefficient K_c, the average effects of soil wetting are incorporated into a general mean K_c. Some guidelines for the estimation of $K_{c\ adj}$ are given in the following sections. The single crop coefficient can also be derived from the adjusted K_{cb} by considering the frequency of soil wetting, i.e., during the midseason period, $K_{c\ adj} = K_{cb\ adj} + 0.05$ for infrequent wetting and $K_{cb\ adj} + 0.10$ for wettings of up to once a week.

For more frequent wettings, the dual crop coefficient approach should be used.

Alternatively, Equations can be used to determine K_c instead of K_{cb}. Then, $K_{c\ min}$ in Equations can be set equal to $K_{c\ ini}$, where $K_{c\ ini}$ is estimated. The use of $K_{c\ ini}$ incorporates effects of soil evaporation and therefore serves as a lower limit on the estimate for $K_{c\ mid}$.

Water Stress Conditions

Where rainfall or irrigation is low, water stress might be induced and the evapotranspiration will drop below the standard crop evapotranspiration, ET_c. The reduction in the value for K_c under conditions of low soil water availability is determined using the stress coefficient.

MID-SEASON STAGE - ADJUSTMENT FOR SPARSE VEGETATION

Adjustment from Simple Field Observations

As a rough approximation for K_c during the mid-season stage for crops that usually nearly completely shade the soil under pristine conditions, but where cover is reduced due to disease, stress, pests, or planting density, the values for $K_{c\ mid}$ and $K_{cb\ mid}$ can be reduced by a factor depending on the actual vegetation development:

$$K_{c\ adj} = K_c - A_{cm}$$

where, K_c the K_c from table ($K_{c\ mid}$) or 17 ($K_{cb\ mid}$) after adjusting it for climate,

$K_{c\ adj}$ the adjusted K_c ($K_{c\ mid\ adj}$ or $K_{cb\ mid\ adj}$).

The K_c adjustment using Equation does not apply when crops are frequently wetted and increased soil evaporation compensates for the reduced ground cover. Under these conditions Equation should be applied only to K_{cb}.

The adjustment coefficient, A_{cm}, is estimated from:

$$A_{cm} = 1 - \left[\frac{LAI}{LAI_{dense}}\right]^{0.5}$$

Where, LAI is the actual leaf area index and LAI_{dense} is the leaf area index expected for the same crop under normal, standard crop management practices. The values for LAI in the above equation can be replaced by values for the ground cover fraction (f_c):

$$A_{cm} = 1 - \left[\frac{f_c}{f_{cdense}}\right]^{0.5}$$

Example. First approximation of the crop coefficient for the mid-season stage for sparse vegetation.

A tomato crop was grown at Davis, California, United States in 1980 and only developed 50% ground cover during the midseason period. The height of the tomato crop was 0.75 m. The typical percentage of ground cover for tomatoes at effective full cover at Davis is 85 to 90% and corresponds to the $K_{cb\ mid}$ for tomatoes. The mean values for wind speed and minimum relative humidity during the midseason period were u_2 = 1.1 m/s and RH_{min} = 30%. The latitude at Davis is 38.5° N and the midpoint of the midseason occurs on July 20. What is an adjusted $K_{cb\ mid}$ for tomatoes that reflects the 50% ground cover condition?

From tables and 17, $K_{c\ mid}$ = 1.2 and $K_{cb\ mid}$ = 1.15.

Following adjustments for climate where u_2 = 1.1 m/s, RH_{min} = 30% and mean crop height = 0.75 m,

$K_c = K_{c,\ table} + [0.04(1.1 - 2) - 0.004(30 - 45)](0.75/3)^{0.30}$
$= K_{c,\ table} + 0.02$ yields,
$K_{c\ mid} = 1.22$ and $K_{cb\ mid} = 1.17$.

The ground cover fraction implied in the tabulated values for tomatoes grown under pristine conditions is about 85% ($f_{c\ dense} = 0.85$). For a sparse tomato crop where $f_c = 0.50$,

From Equation

$$A_{cm} = 1 - (0.50/0.85)^{0.5} = 0.23$$

The $K_{cb\ mid\ adj}$ and $K_{c\ mid\ adj}$ for 50% ground cover is

$$K_{cb\ mid\ adj} = 1.17 - 0.23 = 0.94$$

$$K_{c\ mid\ adj} = 1.22 - 0.23 = 0.99$$

Compare the results with Example where a more precise derivation of $K_{cb\ mid\ adj}$ is made.

As a first estimate, the crop coefficient is expected to be 20% lower than the value under pristine conditions.

ESTIMATION OF $K_{CB\ MID}$ FROM LEAF AREA INDEX (LAI)

Natural vegetation typically has less leaf area or fraction of ground cover than does agricultural vegetation that has been developed for full ground cover and for soil water conditions favouring vigorous growth. This is especially true in semi-arid and arid climates.

The value for $K_{cb\ mid}$ for natural or non-pristine vegetation should be reduced when plant density and/or leaf area are lower than for full cover conditions (generally defined as when LAI ≥).

Where LAI can be measured or approximated, a peak $K_{cb\ mid}$ for natural, non-typical or non-pristine agricultural vegetation can be approximated similar to a procedure used by Ritchie as:

$$K_{cb\ mid} = K_{c\ min} + (K_{cb\ full} - K_{c\ min})(1 - \exp[-0.7\ LAI])$$

where, $K_{cb\ mid}$ estimated basal K_{cb} during the mid-season when plant density and/or leaf area are lower than for full cover conditions,

$K_{cb\ full}$ estimated basal K_{cb} during the mid-season (at peak plant size or height) for vegetation having full ground cover or LAI > 3,

$K_{c\ min}$ the minimum K_c for bare soil ($K_{c\ min} \approx 0.15 - 0.20$),

LAI actual leaf area index, defined as the area of leaves per area of underlying ground surface averaged over a large area. Only one side of leaves is counted [$m^2\ m^{-2}$].

Equation is recommended for annual types of vegetation that are either natural or are in a non-pristine state due to sparse density or effects of some type of environmental stress on growth.

The relationship expressed in Equation produces results similar to those suggested by Ritchie (1974). For LAI > 3, $K_{cb\ mid} \approx K_{cb\ full}$. The

LAI used in equation should be the 'green' LAI representing only healthy leaves that are active in vapour transfer.

Measuring and Estimating LAI

LAI can be measured directly by harvesting all green healthy leaves from vegetation over a measured or prescribed area, for example, 1 m^2 or 10 m^2, and then measuring and summing the areas of individual leaves using photometric methods or by measuring areas of several representative leaves, averaging, and then multiplying by the total number of leaves counted.

In the absence of measurements for LAI, LAI can be estimated for sparse, annual vegetation as:

where, LAI_{dense} LAI for the particular plant species under normal, 'dense' or pristine growing conditions. LAI_{dense} can be obtained from various physiological sources and textbooks on crops and vegetation.

Population number of plants per unit area of soil surface under the actual growing conditions [No. m^{-2}].

$Population_{dense}$ number of plants per unit area of soil surface under the 'dense' or pristine growing conditions [No. m^{-2}].

a a = 0.5 when population is formed from vigorous growing plants; a = 1 when plants are less vigorous.

The 0.5 exponent in the equation simulates the tendency for vegetation to compensate for reduced stand density by increasing the size and total leaf areas for individual plants. Therefore, LAI does not fall in direct proportion to plant population.

Under conditions where the plant size does not increase with reduced stand density, the 'a' exponent in the equation should be set to 1 (a = 1). These latter conditions may occur where soil fertility is poor or where soil salinity, soil water stress, or waterlogging inhibit both growth and stand density, so that the growth of individual plants is retarded.

Estimation of $K_{cb\ mid}$ from Effective Ground Cover ($f_{c\ eff}$)

Where only estimates of the fraction of soil surface effectively covered by vegetation are available, the following approximation for $K_{cb\ mid\ adj}$ can be used:

$$K_{cbmid} = K_{c\min} + (K_{cbfull} - K_{c\min})\left(\min\left(1, 2f_{,}(f_{ceff})\left(\frac{1}{1+h}\right)\right)\right)$$

where, $K_{cb\ mid}$ estimated basal K_{cb} during the mid-season when plant density and/or leaf area are lower than for full cover conditions,

$K_{cb\ full}$ estimated basal K_{cb} during the mid-season (at peak plant size or height) for vegetation having full ground cover or LAI > 3,

$K_{c\,min}$ the minimum K_c for bare soil (in the presence of vegetation) ($K_{c\,min} \approx 0.15 - 0.20$),

f_c observed fraction of soil surface that is covered by vegetation as observed from nadir (overhead) [0.01 – 1],

$f_{c\,eff}$ the effective fraction of soil surface covered or shaded by vegetation [0.01 – 1],

h the plant height [m].

Stomatal conductance and water transport within plants may limit ET under conditions of sparse, tall vegetation. Under these conditions, $K_{cb\,mid}$ is limited by the "$2f_c$" term in Equation. Equation applies well to trees and shrubs.

Measuring and Estimating $f_{c\,eff}$

$f_{c\,eff}$ should normally represent the fraction of the soil surface that is shaded by vegetation. This value is generally larger than f_c, the actual fraction of the soil surface that is covered by vegetation as observed from directly overhead.

The amount of shading represents the amount of solar radiation intercepted by plants for potential conversion into evapotranspiration. The total fraction of shading is a function of the sun angle and the horizontal size and shape of individual plants (or rows) relative to their height.

$f_{c\,eff}$ for 'rectangular' shaped canopies such as most agricultural plant rows can be approximated as:

$$f_{ceff} = f_c \left[1 + \frac{HWR}{\tan(\eta)} \right] \leq 1$$

where, f_c observed fraction of soil surface that is covered by vegetation as observed from nadir (overhead) [0.01 – 1],

HWR height to width ratio of individual plants or groups of plants when viewed from the east or from the west,

tan(h) tangent of the mean angle of the sun, h, above the horizon during the period of maximum evapotranspiration (generally between 11.00 and 15.00 hours). For most applications, h can be computed at solar noon (12.00 hours).

HWR is computed as:

$$HWR = \frac{h_{canopy}(COS(\Gamma))}{Width}$$

where

h_{canopy} mean vertical height of the canopy area of the plant [m],

Width mean width of the canopy of a plant or group of plants (e.g., row) [m]

Γ angle of plant row from east-west direction [rad] (for east-west rows, Γ= 0; for north-south rows, Γ= π /2)

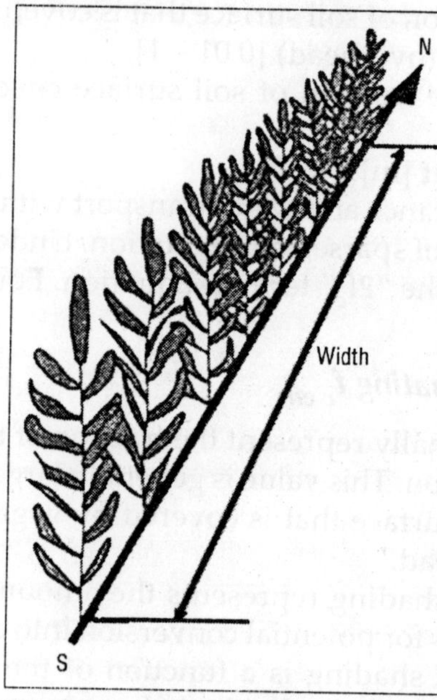

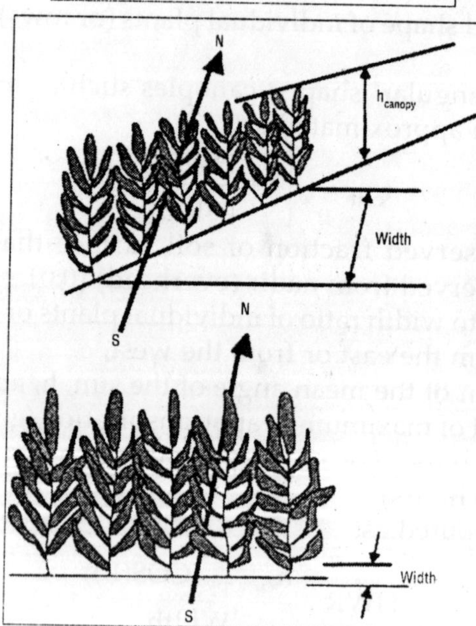

For north-south rows, the HWR would be zero, as cos(π /2) = 0. This implies that rows of plants that run from north to south would have $f_{c\ eff} \approx f_c$ as all soil is exposed to the sun at various times of the day and as the shaded area is the same as the fraction of vegetation cover at midday. For trees or vegetation that do not have canopies that extend to the ground, h_{canopy} does not include the lower trunk length, but only the active canopy. Therefore, in these situations, $h_{canopy} < h$ where h is mean plant height.

For round or spherical shaped canopies such as trees, $f_{c\,eff}$ can be estimated as:

$$f_{ceff} = \frac{f_c}{\sin(\eta)} \leq 1$$

where sin(η) is the sine of the mean angle of the sun, η, above the horizon during the period of maximum evapotranspiration (generally between 11.00 and 15.00)

Mean angle of the sun above the horizon during the period of maximum evapotranspiration (η)

The sine of h can be calculated for any specific time of day as:

$$\sin(\eta) = \sin(\varphi)\sin(\delta) + \cos(\varphi)\cos(\delta)\cos(\omega)$$

where, φ latitude [rad] (negative for southern latitudes)

δ solar declination [rad]

ω solar time angle [rad]

Generally, $f_{c\,eff}$ can be calculated at solar noon (12.00), so that ω = 0 and the above equation reduces to:

$$\sin(\eta) = \sin(\varphi)\sin(\delta) + \cos(\varphi)\cos(\delta)$$

The value for h can be obtained by taking the arcsine of the above equation.

Estimation of $K_{cb\,full}$

Agricultural Crops

Non-pristine agricultural crops represent crops that have not developed to their potential due to environmental stresses caused by soil water shortage, fertility, disease, grazing or insect damage or due to low plant density. The value for $K_{cb\,full}$ in Equations can be taken as the $K_{cb\,mid}$ value listed for any "full-cover" crop ($f_{c\,eff}$ ~ 1), after adjusting it for climate

$$K_{cbfull} = K_{cbmidTable} + [0.04(u_2 - 2) - 0.004(RH_{min} - 45)]\left(\frac{h}{3}\right)^{0.3}$$

where, u_2 mean value for wind speed at 2 m height during the mid-season [m s^{-1}],

RH_{min} mean value for minimum daily relative humidity during the mid-season [%].

h mean maximum plant height [m].

Natural Vegetation and Crops

For natural vegetation, nonfull-cover crops, or for crops, $K_{cb\ full}$ can be approximated as a function of climate and mean plant height for areas of vegetation that are greater than a few hectares:

$$K_{cbfull} = K_{cbh} + [0.04(u_2 - 2) - 0.004(RH_{min} - 45)]\left(\frac{h}{3}\right)^{0.3}$$

where, $K_{cb,\ h}$ $K_{cb\ mid}$ for full cover vegetation (LAI > 3) under sub-humid and calm wind conditions (RH_{min} = 45% and u_2 = 2 m s^{-1}),

u_2 mean value for wind speed at 2 m height during the mid-season [m s^{-1}],

RH_{min} mean value for minimum daily relative humidity during the mid-season [%]

h mean maximum plant height [m].

The value for $K_{cb,\ h}$ is estimated as:

$K_{cb,\ h}$ = 1.0+0.1 h for h ≤ 2 m (101)

where $K_{cb,\ h}$ is limited to ≤ 1.20 when h > 2 m. The value of 1.2 represents a general upper limit on $K_{cb\ mid}$ for tall vegetation having full ground cover and LAI > 3 under the sub-humid and calm wind conditions.

This limit of 1.2 is adjusted for other climatic conditions in Equation. Equations 100 and 101 produce a general approximation for the increase in $K_{cb\ full}$ with plant height and climate. The form of these equations adheres to trends represented in Equation.

For small, isolated stand sizes, $K_{cb\ full}$ may need to be increased beyond the value given by Equation.

Equations can be used to estimate or to reduce K_{cb} for non-pristine agricultural vegetation. The exponents in Equations reflect the effects of microscale advection (transfer) of sensible heat from dry soil surfaces between plants toward plant leaves, thereby increasing ET per unit leaf area, and the effects of increased aerodynamic roughness as the value for LAI decreases.

Equation suggests that as h increases, total leaf area and effective roughness of vegetation increase, thereby increasing the crop coefficient.

In addition, as h increases, more opportunity for microadvection of heat from soil to canopy occurs and turbulent exchange within the canopy increases for the same amount of ground coverage. All of these factors affect the relative magnitude of $K_{cb\ mid}$.

Equations should be used with caution as they provide only an estimate of the maximum K_{cb} expected during peak plant growth for vegetation with healthy transpiring leaves and a dry soil surface.

Where stomatal control is greater than for typical agricultural vegetation, then the K_{cb} should be further reduced using the recommendations set out in the next section.

Estimation of Mid-season Crop Coefficient

Estimate $K_{cb\ mid}$ and $K_{c\ mid}$ for rectangular shaped 2 m tall vegetation that is as tall as it is wide, where 30% of the ground is covered by vegetation ($f_c = 0.3$) on 19 July (day 200 of the year) and at latitude 40°N. $RH_{min} = 55\%$ and $u_2 = 1.5$ m/s during the mid-season period.

On day J = 200 at latitude ($\varphi = 40 (\pi /180) = 0.70$ radians (40°N), from Equation, the solar declination $\delta = 0.36$ radians.

At solar noon ($\omega = 0$):

$$\sin(\eta) = \sin(\varphi)\sin(\delta) + \cos(\varphi)\cos(\delta) = 0.94$$

The value for η by taking the arcsine of above value is 1.24 radians and $\tan(\eta) = 2.8$. If $f_c = 0.3$ and the HWR for the vegetation is 1, then $f_{c\ eff}$ from is: $0.3(1 + 1/2.8) = 0.41$.

From Equation $K_{cb,\ h} = 1.0 + 0.1(2) = 1.20$ (≤ 1.20, so OK)

From Equation $K_{cb\ full} = 1.20 + (0.04(1.5 - 2) - 0.004(55 - 45))\ (2/3)^{0.3}$ $=1.15$

Therefore, $K_{cb\ mid}$ estimated by Equation for $K_{c\ min} = 0.15$ is

$$K_{cb\ mid} = K_{c\ min} + (K_{cb\ full} - K_{c\ min})\ \min[1,\ 2f_c,\ (f_{c\ eff})^{(1/(1+h))}]$$

$$= 0.15 + (1.15 - 0.15)\ \min[1, 2(0.3), (0.4)^{(1/(1+2))}] = 0.75$$

This value does not need any further adjustment for climate.

$K_{c\ mid}$ (where $K_{c\ mid}$ includes average wetting effects) can be derived from $K_{cb\ mid}$ using the guidelines presented in the calculation procedures at the beginning.

$$K_{c\ mid} = K_{cb\ mid} + (0.05...\ 0.10) = 0.80..\ 0.85$$

depending on the frequency of soil wetting.

The estimated crop coefficients for the mid-season stage are

$$K_{cb\ mid} = 0.75 \text{ and } K_{c\ mid} = 0.80 \text{ to } 0.85.$$

Estimation of Mid-season Crop Coefficient for Reduced Ground Cover

A more precise estimate of $K_{cb\ mid}$ for the tomato crop of Example that only developed 50% ground cover at Davis, California, the United States can be calculated if one knows that the tomato crop was planted in 1.52 m rows running east-west, that the crop reached a plant height of 0.75 m and that the height to width ratio of the tomato crop can be estimated as about 1.0 for the east-west rows. The latitude is 38.5°N and the midpoint of the mid-season occurs on July 20.

What is the adjusted $K_{cb\ mid}$ for tomatoes that reflects the 50%

ground cover condition On day J = 201 (20 July) at latitude $\varphi = 38.5$ (π /180) = 0.67 radians (38.5°N), from Equation the solar declination δ = 0.36 radians. At solar noon ($\omega = 0$):

$$\sin(\eta) = \sin(\varphi)\sin(\delta) + \cos(\varphi)\cos(\delta) = 0.95$$

The value for η by taking the arcsine of the above value is 1.26 radians. Therefore, for the observed HWR = 1 and $f_c = 0.5$, the effective soil cover for the east-west rows can be calculated as:

$$f_{c\,eff} = 0.5\,[1 + 1/\tan(1.26)] = 0.66$$

The $K_{cb\,mid}$ in Table representing a full cover tomato crop is 1.15 and the average h for fully developed tomatoes (this variety) is about 0.75 also. Following adjustment for climate

$$K_{cb\,full} = 1.15 + [0.04(1.1 - 2) - 0.004(30 - 45)]\,(0.75/3)^{0.30} = 1.17$$

From Equation and using $K_{c\,min} = 0.15$, the adjusted $K_{cb\,mid}$ for 50% ground cover would be:

$$K_{cb\,mid\,adj} = 0.15 + (1.17 - 0.15)\,\min(1, 2(0.5), 0.66^{1/(1+0.76)}) = 0.95$$

The results $K_{cb\,mid\,adj} = 0.95$ for 50% ground cover are similar to the first estimate calculated in Examples compare with the measured $K_{cb\,mid}$ ~ 0.90 to 1.00 as determined by precision lysimeter by Pruitt *et al.*

Mid-Season Stage - Adjustment for Stomatal Control

The value for $K_{cb\,full}$ in Equations may need to be reduced for vegetation that has a high degree of stomatal control. For vegetation such as some types of desert vegetation or trees with leaf resistance significantly greater than that of most agricultural crops where r_l is commonly about 100 s m^{-1}, the $K_{cb\,mid}$ estimated using Equations can be modified by multiplying by a resistance correction factor, F_r. The resistance correction factor is developed based on the FAO Penman-Monteith equation:

$$F_r \approx \frac{\Delta + \gamma(1 + 0.34u_2)}{\Delta + \gamma\left(1 + 0.34u_2 \frac{r_l}{100}\right)}$$

where, r_l mean leaf resistance for the vegetation in question [s m^{-1}]

The mean leaf resistance r_l is 100 s m^{-1} for the grass ET_o reference and for many agricultural crops.

Values for r_l for many agricultural and non-agricultural plants can be found in Korner *et al.* and Allen *et al.*

Equation reflects the fixed aerodynamic roughness of grass rather than the roughness of the specific vegetation, since the adjusted K_c is multiplied by the grass ET_o and the K_c already reflects the effects of the aerodynamic roughness for the specific vegetation.

Estimation of Kcb Mid from Ground Cover with Reduction for Stomatal Control

A grove of olive trees has a tree spacing of 10 m. The horizontal diameter of the trees as viewed from overhead is 5 m. The tree height is 5 m. The lower 1.5 m of the trees have no foliage. The ground cover between the trees is bare. The mean u_2 during the mid-season growth stage is 2 m/s and the mean RH_{min} = 25%. The midpoint of the mid-season growth stage is on 29 June (i.e., day 180 of the year). The latitude of the location is 30°N.

Estimate $K_{cb\ mid}$ using Equation for the 10x10 m and for a 5 × 10 m spacing.

On day J = 180 (29 June) at latitude (φ = 30 (π /180) = 0.52 radians (30°N) and from Equation the solar declination δ = 0.405 radians. At solar noon (ω – 0):

$$\sin(\eta) = \sin(\varphi)\sin(\delta) + \cos(\varphi)\cos(\delta) = 0.99$$

As olive trees have somewhat round shapes, the effective fraction of ground cover can be estimated as $f_{c\ eff} - f_c/(\sin(h))$.

f_c = area of canopy/area tree spacing = (p $(5)^{2/4}$)/(10)(10) = 0.196

$$f_{c\ eff} = 0.196/0.99 = 0.20$$

From Equation: $K_{cb,\ h}$ = 1.0 + 0.1(5) > 1.2 or $K_{cb,\ h}$ = 1.2

From Equation: $K_{cb\ full}$ = 1.2 + [0 – 0.004(25 – 45)]$(5/3)^{0.3}$ = 1.29

From Equation and using $K_{c\ min}$ = 0.15: $K_{cb\ mid}$
= 0.15 + (1.29 – 0.15) min(1,2(0.196), $(0.20)^{1/(1+5)}$) = 0.60

Körner *et. al.* indicate that olives (*Olea europaea*) have r_l of about 420 s/m. Therefore, assuming that average

$$T_{mean} = 25°C$$

and that the elevation of the grove is 0 m (sea level), so that

$$D = 0.189 \text{ kPa}$$

and

$$g = 0.0676 \text{ kPa},$$

F_r is estimated from Equation as:

$$F_r \approx \frac{0.189 + 0.0676(1+0.34(2))}{0.189+0.0676\left(1+0.34(2)\frac{420}{100}\right)} = 0.67$$

The $K_{cb\ mid}$ adjusted for increased stomatal control using F_r is then

$K_{cb\ mid\ adj} = F_r\, K_{cb\ mid}$ = 0.67 (0.60) = 0.40

The value $K_{cb\ mid\ adj}$ = 0.40 estimated for f_c = 0.20 is less than the value for $K_{cb\ mid}$ in Table for olives for f_c - 0.40 to 0.67, due to the differences in f_c. The value from Table is 0.70, which after adjustment for climate using Equation .

If the olives had been planted on a 5 × 10 m spacing, as is common in California, the United States, and which is reflected in the K_{cb} values for olives in Table, then $f_c = 0.39$, $f_{c\,eff} = 0.40$, and $K_{cb\,mid}$ from Equation is $K_{cb\,mid} = 1.04$, so that the estimated $K_{cb\,mid}$ adjusted for stomatal control using $F_r = 0.67$ is $K_{cb\,mid\,adj} = 0.67(1.04) = 0.70$. This value compares with the value of 0.79 obtained from Table for mature trees, after adjustment for climate.

The basal crop coefficient, $K_{cb\,mid}$, taking the low density, climatic condition and stomatal control into account is 0.40. It increases to 0.70 for the 5 × 10 m spacing.

The equation would underestimate F_r (overestimate the reduction in K_{cb}) if used with the actual roughness of the vegetation when $r_l > 100$ s m^{-1} because of the lack in Equation 102 of feedback effects that reduced ET_c has on temperature and vapour pressure deficit profiles over the crop. These parameters generally increase with decreasing ET_c and therefore dampen the reduction in ET_c.

Late Season Stage

During the late season stage, the K_{cb} begins to decrease until it reaches $K_{cb\,end}$ at the end of the growing period. Values for $K_{cb\,end}$ can be scaled from $K_{cb\,mid}$ in proportion to the health and leaf condition of the vegetation at termination of the growing season and according to the length of the late season period (i.e., whether leaves senesce slowly or are killed by frost). Values for $K_{c\,end}$ can be similarly scaled from $K_{c\,mid}$; however, the reduction in $K_{c\,end}$ will be affected by the frequency of wetting by irrigation or precipitation and $K_{c\,end}$ may be proportionally less.

If estimated from Equations, $K_{cb\,end}$ should be reduced if it is to represent K_c values for plants with stomatal control that is greater than that for agricultural vegetation (where $r_l \approx 100$ s m^{-1}) or to reflect effects of ageing and senescence on stomatal control. In these situations, the estimated $K_{cb\,end}$ values should be multiplied by the F_r. Alternatively, they can be reduced by about 10% for each doubling of r_l above 100 s m^{-1} when mean daily air temperature (T_{mean}) is about 30° C and by about 20% for each doubling of r_l above 100 s m^{-1} when T_{mean} is about 15° C.

Alternatively, the value for $K_{cb\,end}$ can be reduced relative to the calculated value for $K_{cb\,mid}$ in proportion to the fraction of green healthy leaves remaining at the end of the late season stage relative to that during the mid-season. This can often be based on a visual survey of me field and may therefore be a subjective observation.

The f_c parameter and h are probably the simplest indices to estimate in the field.

Estimating $ET_{c\ adj}$ using Crop Yields

A simple, linear crop-water production function was introduced in the FAO Irrigation and Drainage to predict the reduction in crop yield when crop stress is caused by a shortage of soil water. This function was presented earlier as Equation:

$$\left(1-\frac{Y_a}{Y_m}\right)=K_y\left(1-\frac{ET_{cadj}}{ET_c}\right)$$

where, Y_a = actual yield of the crop [kg ha^{-1}]

Y_m = maximum (expected) yield in absence of environmental or water stresses

K_y = yield response factor.

ET_c = potential (expected) crop evapotranspiration in the absence of environmental or water stresses (K_c ET_o)

$ET_{c\ adj}$ = actual (adjusted) crop evapotranspiration as a result of environmental or water stresses

Values for K_y have been reported in Paper for a wide range of crops for predicting the effect of water stress and associated reduction in $ET_{c\ adj}$ on crop yield. Factors are presented there for predicting yield reductions for when stress occurs in only one crop growth stage, or when stress is distributed throughout the growing period.

Many environmental stresses such as water shortage, salinity, low fertility and disease impact yield by reducing the amount of $ET_{c\ adj}$ relative to the potential amount ET_c. The same can be true for when yields are reduced due to the use of low densities for plant populations. Therefore, for very general estimates of $ET_{c\ adj}$, one can invert Equation and solve for the stress factor, K_s:

$$K_s=1-\frac{1}{K_y}\left(1-\frac{Y_a}{Y_m}\right)$$

where K_s is multiplied by K_{cb} or by K_c in equations to predict the $ET_{c\ adj}$ in the presense of the water or other environmental stresses or for low plant populations or virility. The $ET_{c\ adj}$ predicted using K_s from equation provides only a very general and approximate estimate of monthly or even seasonal evapotranspiration. Equation 103 works best for forage or other indiscriminate crops where the value for K_y is relatively constant during the season.

Equation is generally only valid for use in predicting actual crop evapotranspiration for use in regional water balance studies, for studies of ground-water depletion and recharge, or for estimating historical water use. The procedure is not valid for predicting ET_c for daily or weekly time periods due to the very general nature of the K_y coefficient and the seasonal time scale of the crop yield. The procedures presented previously for adjusting ET_c using a daily soil water balance, salinity functions, or reductions in K_c based on leaf area or fraction of ground cover are recommended over the use of equation.

Approximate Estimation of Ks from Crop Yield Data

An irrigation scheme (project) cultivates dry, edible beans. There is known to be a shortage of irrigation water and a corresponding reduction in crop yield. The reported yield for the scheme averages 1100 kg/ha. The potential yield for the region and variety of beans, in the absence of water or environmental stresses and with good soil fertility is 1800 kg/ha.

From FAO Irrigation and Drainage Paper, the K_y for dry beans, assuming that stresses are distributed uniformly through the growing season, is 1.15. Therefore, from Equation, the estimated K_s to apply with Equation 80 for the growing season is:

$$K_s = 1 - \frac{1}{1.15}\left(1 - \frac{1100}{1800}\right) = 0.66$$

Therefore, the $ET_{c\ adj}$ for the season is predicted to be only 0.66 of maximum ET_c under pristine growing conditions.

The estimated seasonal $ET_{c\ adj}$ is predicted to be $ET_{c\ adj} = 0.66\ ET_c$ where ET_c is predicted as $ET_c = K_c\ ET_o$.

Chapter 9

Management Practices for Vegetation

This chapter discusses various types of factors that may cause the values for K_c and ET_c to deviate from the standard values described. These factors refer to the effects of surface mulches, intercropping, small areas of vegetation and specific cultivation practices.

This chapter is intended to serve as a resource for situations where cultivation practices are known to deviate from those assumed, but where estimates of K_c and ET_c are still necessary. This chapter is by no means exhaustive. The intent is to demonstrate some of the procedures that can be used to make adjustments to K_c to account for deviations from standard conditions.

EFFECTS OF SURFACE MULCHES

Mulches are frequently used in vegetable production to reduce evaporation losses from the soil surface, to accelerate crop development in cool climates by increasing soil temperature, to reduce erosion, or to assist in weed control. Mulches may be composed of organic plant materials or they may be synthetic mulches consisting of plastic sheets.

Plastic Mulches

Plastic mulches generally consist of thin sheets of polyethylene or a similar material placed over the ground surface, especially along the plant rows. Holes are cut into the plastic film at plant spacings to allow the plant vegetation to emerge. Plastic mulches can be transparent, white or black. Colour influences albedo mainly during the early stages of the crop. However, as the intention is to use a simple procedure for adjusting K_c for mulched crops, no distinction is made between the different types of plastic mulches.

Plastic mulches substantially reduce the evaporation of water from the soil surface, especially under trickle irrigation systems. Associated with the reduction in evaporation is a general increase in transpiration from vegetation caused by the transfer of both sensible and radiative heat from the surface of the plastic cover to adjacent vegetation.

Even though the transpiration rates under mulch may increase by an average of 10-30% over the season as compared to using no mulch, the K_c values decrease by an average of 10-30% due to the 50-80% reduction in soil evaporation. A summary of observed reductions in K_c, in evaporation, and increases in transpiration over growing seasons is given in Table for five horticultural crops. Generally, crop growth rates and vegetable yields are increased by the use of plastic mulches.

Table. Approximate reductions in K_c and surface evaporation and increases in transpiration for various horticultural crops under complete plastic mulch as compared with no mulch using trickle irrigation.

Crop	Reduction [1] in Kc (%)	Reduction [1] in evaporation (%)	Increase [1] in transpiration (%)	Source
Squash	5-15	40-70	10-30	Safadi (1991)
Cucumber	15-20	40-60	15-30	Safadi (1991)
Cantaloupe	5-10	80	35	Battikhi and Hill (1988)
Watermelon	25-30	90	-10	Battikhi and Hill (1986), Ghawi and Battikhi (1986)
Tomato	35	not reported	not reported	Haddadin and Ghawi (1983)
Average	10-30	50-80	10-30	

Single Crop Coefficient, Kc

To consider the effects of plastic mulch on ET_c, the values for $K_{c\ mid}$ and $K_{c\ end}$ for the horticultural crops listed in Table can be reduced by 10-30%, depending on the frequency of irrigation (use the higher value for frequent trickle irrigation).

The value for $K_{c\ ini}$ under mulch is often as low as 0.10. When the plastic mulch does not entirely cover the soil wetted by the drip emitters, or where substantial rainfall occurs, then the reduction in $K_{c\ mid}$ or $K_{c\ end}$ will be less, in proportion to the fraction of wet surface covered by the mulch.

Dual Crop Coefficient, Kcb + Ke

When estimating basal K_{cb} for mulched crops, less adjustment is normally needed to the K_{cb} curve, being of the order of a 5-15% reduction in K_{cb}, as it is generally understood that the 'basal' evaporation of water from the soil surface is less with a plastic mulch, though the transpiration is increased.

The effect on K_{cb} could be greater in some situations and with some types of low density crops. Local calibration of K_{cb} (and K_c) for use with mulch culture is encouraged.

When calculating the soil evaporation coefficient K_e with plastic mulch, the f_w should represent the relative equivalent fraction of the ground surface that can contribute to evaporation through the vent holes in the plastic cover and to the fraction of surface that is wetted, but is not covered by the mulch. The effective area of vent holes is normally two to four times the physical area of the vents (or even higher) to account for vapour transfer from under the sheet.

Organic Mulches

Organic mulches are often used with orchard production and with row crops under reduced tillage operations. Organic mulches may consist of unincorporated plant residues or foreign material imported to the field such as straw.

The depth of the organic mulch and the fraction of the soil surface covered can vary widely. These two parameters will affect the amount of reduction in evaporation from the soil surface.

Effects of Surface Mulch

A plastic mulch is placed over cucumbers under drip irrigation. The mulch is clear plastic covering the entire field surface, with small openings at each plant. Adjust both the mean and basal K_c values for this crop to reflect the presence of the mulch.

From table, $K_{c\ ini}$, $K_{c\ mid}$ and $K_{c\ end}$ for fresh market cucumbers have values equal to 0.4, 1.0 and 0.75.

As the plastic mulch is continuous with only small vents at each plant, the $K_{c\ ini}$ is assumed to be only 0.10 (this value should be adjusted upward if precipitation occurs).

The $K_{c\ mid}$ and $K_{c\ end}$ values are estimated as:

$$K_{c\ mid} = 0.85\ (1.0) = 0.85$$

$$K_{c\ end} = 0.85\ (0.75) = 0.64$$

where the 0.85 multipliers are derived from table and reflect an assumed 15% reduction in ET_c due to the mulch, assuming an approximately weekly irrigation frequency.

From table, the values for $K_{cb\ ini}$, $K_{cb\ mid}$, and $K_{cb\ end}$ are 0.15, 0.95 and 0.7 for this same cucumber crop. The $K_{cb\ ini}$ is assumed to be similar to the $K_{c\ ini}$ for mulched cover and is therefore set equal to 0.10. The $K_{cb\ mid}$ and $K_{cb\ end}$ values are estimated to be reduced by 10% so that:

$$K_{cb\ mid} = 0.9\ (0.95) = 0.86$$
$$K_{cb\ end} = 0.9\ (0.7) = 0.63$$

These basal values are similar to the adjusted values for K_c. This is expected as evaporation from the mulch covered surface can be ignored. Additional adjustment to these K_c values to account for climate is necessary using Equation .

The values for mean K_c and K_{cb} are similar with values of 0.10 for the initial stage, 0.85 for the mid-season stage and 0.64 at the end of the late season stage.

Single Crop Coefficient, K_c

A general rule when applying K_c from Table is to reduce the amount of soil water evaporation by about 5% for each 10% of soil surface that is effectively covered by an organic mulch.

For example, if 50% of the soil surface were covered by an organic crop residue mulch, then the soil evaporation would be reduced by about 25%.

- In the case of $K_{c\ ini}$, which represents mostly evaporation from soil, one would reduce $K_{c\ ini}$ by about 25% in this situation.
- In the cases of $K_{c\ mid}$ and $K_{c\ end}$, one would reduce these values by 25% of the difference between ($K_{c\ mid}$ - $K_{cb\ mid}$) and ($K_{c\ end}$ - $K_{cb\ end}$) from Tables. Generally, the differences between values in Tables are only 5-10% so that the adjustment to $K_{c\ mid}$ and $K_{c\ end}$ to account for an organic mulch may not be very large.

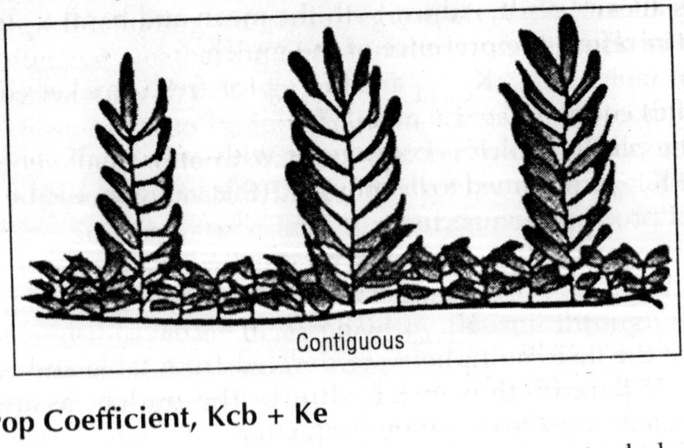

Contiguous

Dual Crop Coefficient, Kcb + Ke

When applying the approach with a separate water balance of the surface soil layer, the magnitude of the evaporation component (K_e ET_o) should be reduced by about 5% for each 10% of soil surface covered by the organic mulch. K_{cb} is not changed.

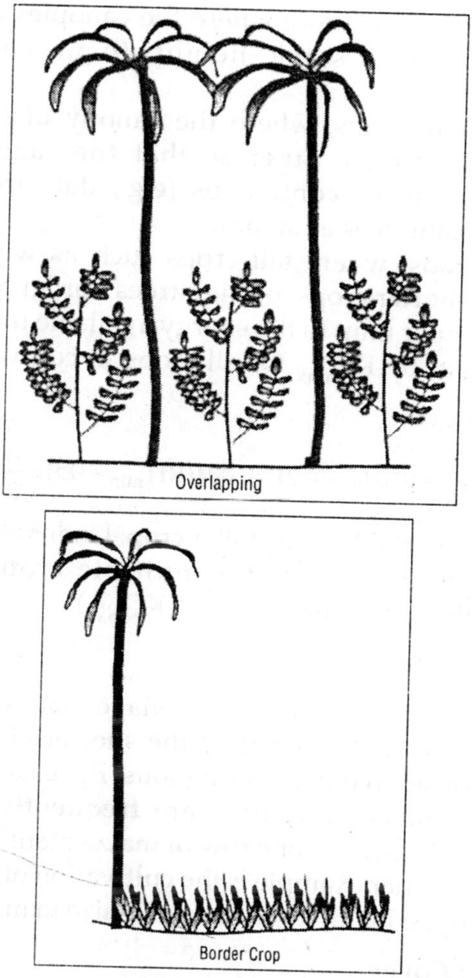

Fig. Different Situations of Intercropping

These recommendations are only approximate and attempt to account for the effects of partial reflection of solar radiation from residue, microadvection of heat from residue into the soil, lateral movement of soil water from below residue to exposed soil, and the insulating effect of the organic cover. As these parameters can vary widely, local observations and measurements are required if precise estimates are required.

INTERCROPPING

Intercropping refers to the situation where two different crops are grown together within one field. For the estimation of the crop coefficient, a distinction is made between:

- Contiguous vegetation, where the canopies of the two crops intermingle at some height (e.g., corn and beans intercropping).
- Overlapping crops, where the canopy of one crop is well above that of the other so that the canopies cannot be considered to be contiguous (e.g., date trees overlapping pomegranate trees at an oasis).
- Border crops, where tall crops such as windbreaks border fields of shorter crops, or high trees border a field crop.

There is an upper limit to the energy available to evaporate water. This is represented by $K_{c\,max}$ for all crops in cultivated fields larger than 3-5 ha:

$$K_{c\,max} = \max\left(\left\{1.2 + [0.04(u_2 - 2) - 0.004(RH_{min} - 45)]\left(\frac{h}{3}\right)^{0.3}\right\}\{K_{cb} + 0.05\}\right)$$

Where h is the height for the taller crop. Under all conditions when combining crop coefficients for multiple crops, K_c should be constrained by this upper bound ($K_c \le K_{c\,max}$).

Contiguous Vegetation

Where the taller crop has canopy foliage that extends down to the same elevation as that of the top of the shorter crop, the vegetation canopy can be considered to be contiguous. For example, in Africa and South America, maize and beans are frequently intercropped as contiguous vegetation, with one row of maize planted per one or more rows of beans. Another example is the cultivation of five to seven rows of wheat intercropped with three rows of maize in many areas of China.

Similar Ground Cover

Where the leaf area or fraction of ground covered by the vegetation (f_c) is similar for each crop, the K_c in Tables for the taller crop (if this K_c is higher) can be taken to represent the entire field. The taller crop will act in some sense as a clothesline so that K_c (and ET_c) for the taller crop per unit of ground area is increased over that given in Table.

However, the K_c (and ET_c) for the shorter crop will be reduced due to the windbreak effect by the taller crop. As a result, the K_c for the field as a whole may be similar to the weighted average of the K_c values for the two crops from Tables, or, the total K_c may more closely follow the K_c predicted for a field sown entirely to the taller crop ($K_{c\,field} \approx K_{c\,taller\,crop}$). Yields for the shorter crop may be reduced relative to those for single cultivar production due to the effects of shading by the taller crop and the competition for soil water.

Different Ground Cover

Where the fractions of ground covered by each crop are different, the K_c for an intercropped field can be estimated by weighting the K_c values for the individual crops according to the fraction of area covered by each crop and by the height of the crop:

$$K_{cfield} = \frac{f_1 h_1 K_{c1} + f_2 h_2 K_{c2}}{f_1 h_1 + f_2 h_2}$$

where f_1 and f_2 are the fractions of the field surface planted to crops 1 and 2, h_1 and h_2 are the heights of crops 1 and 2, and K_{c1} and K_{c2} are the K_c values for crops 1 and 2.

Overlapping Vegetation

Where intercropping entails overlapping of spacings, the canopy of one crop is well above the other. This is the case, for example in southern California, where citrus trees are planted in date palm groves. Where a normal dense spacing is used for both the dates and for the citrus trees, the K_c may increase as the density of the combined vegetation increases, proportional to the increase in LAI, with maximum K_c constrained by either $K_{c\,max}$ n 72) or by $K_{cb\,full}$ unless the total field area is small so that there is an additional clothesline or oasis effect as discussed in the next section.

Intercropped Maize and Beans

Determine the representative $K_{c\,mid}$ for a situation where a single 1 m wide row of maize is grown for each 2 m of squash, where $RH_{min} \approx 45\%$ and $u_2 \approx 2$ m/s.

The $K_{c\,mid}$ and h for maize is 1.20 and 2 m and the $K_{c\,mid}$ and h for squash is 0.95 and 0.3 m. No correction is needed for climate. The representative $K_{c\,mid}$ can be obtained by weighting the individual $K_{c\,mid}$ values according to the fraction of the field surface allocated to each crop ($f_1 \approx 0.3$ for maize and $f_2 \approx 0.7$ for squash) and according to the heights of the crops:

$$K_{cmid} = \frac{0.30(2)(1.20) + 0.70(0.3)(0.95)}{0.30(2) + 0.70(0.3)} = 0.14$$

Values can be obtained for daily K_c in a similar manner by constructing individual K_c curves and then weighting interpolated values from the individual K_c curves for any specific day.

The crop coefficient for the mid-season and entire field is 1.14.

Border Crops

Where tall crops such as windbreaks or date palms border fields of shorter crops, the upper storey of the taller crop can intercept extra sensible heat energy from the air stream. Under these conditions, the K_c is weighted according to the areas for each crop. However, prior to the weighting, the K_c for the border crop, if taller than the field (interior) crop, should be adjusted for potential clothesline impact.

Small Areas of Vegetation

Natural vegetation and some subsistence agriculture frequently occurs in small groups or stands of plants. The value for K_c for these small stands depends on the type and condition of other vegetation surrounding the small stand. Areas surrounded by vegetation having similar roughness and moisture conditions.

In the majority of cases for natural vegetation or for non-pristine agricultural vegetation, the value for K_c must adhere to upper limits for K_c of approximately 1.20-1.40, when the areal expanse of the vegetation is larger than about 2000 m^2.

This is required as ET from large areas of vegetation is governed by one-dimensional energy exchange principles and by the principle of conservation of energy. ET from small stands (< 2000 m^2) will adhere to these same principles and limits only where the vegetation height, leaf area, and soil water availability are similar to that of the surrounding vegetation.

Overlapping Vegetation

A 20 ha date palm grove in Palm Desert, California, the United States has a tree spacing of 6 m. Interplanted among the rows of palms are small orange trees (50% canopy) on a 6 m spacing. The palm and citrus trees are 3 m from one another in the rows. Height of the palms is 10 m and height of the citrus is 3 m.

The canopy foliage of the palms is well above that of the citrus so that the canopies cannot be considered contiguous. Mean average RH_{min} during the mid-season is 20% and u_2 = 2 m/s. The $K_{c\ mid}$ from Table for dates is 0.95 and when adjusted for humidity and wind using Equation is $K_{c\ mid}$ = 1.09. The $K_{c\ mid}$ from Table for citrus having 50% canopy with no ground cover is 0.60 and when adjusted for humidity and wind using Equation is $K_{c\ mid}$ = 0.70. The interplanting of citrus among the date palms has increased the total leaf area of the orchard so that ET_c for the combined planting (palms and citrus together) will be greater than for either planting alone. The estimated combined $K_{c\ mid}$ will be estimated as a function of the increase in total LAI. First the LAI values of the individual plantings are estimated by inverting Equation to solve for LAI:

$$LAI = -1.4\ln\left[1 - \frac{K_{cb} - K_{c\,min}}{K_{cbfull} - K_{c\,min}}\right]$$

where $K_{c\,min}$ is the minimum basal K_c for bare soil (a 0.15 to 0.20) and $K_{cb\,full}$ is the maximum mid-season K_c expected for the crop if there were complete ground cover, calculated using Equation Based on Equation, with h = 10 m for the date palms and h = 3 m for the citrus, the $K_{cb\,full}$ values for the two crops, assuming complete ground cover for each, are $K_{cb\,full}$ = 1.34 for palms and $K_{cb\,full}$ = 1.30 for citrus (using RH_{min} = 20% and u_2 = 2 m/s). These estimates ignore effects of any unique stomatal control. Therefore, using the above equation, the effective LAI values of the date palms and citrus are estimated to be approximately:

LAI_{palms} = -1.4 ln[1 - (1.09 - 0.15)/(1.34 - 0.15)] = 2.2

LAI_{citrus} = -1.4 ln[1 - (0.70 - 0.15)/(1.30 - 0.15)] = 0.9

Therefore, the effective LAI for the date palm-citrus combination is estimated to be approximately

$$LAI_{combined} = LAI_{palms} + LAI_{citrus} = 2.2 + 0.9 = 3.1.$$

The increase in $K_{c\,mid}$ for the date palm orchard resulting from the increase in LAI from the interplanting of citrus is estimated using a ratio of the LAI-based function introduced in Equation. This results in the relationship:

$$K_{cmid} = K_{cmid\,singlecrop}\left[\frac{1 - \exp(-0.7LAIcombined)}{1 - \exp(-0.7LAIsinglecrop)}\right]$$

where $LAI_{combined}$ is the LAI for the two intercropped plantings combined and $LAI_{single\,crop}$ is the LAI for the taller, single crop. $K_{c\,mid\,single\,crop}$ is the mid-season K_c for the taller, single crop (in this case the date palms). In this application, the above equation is solved as:

$$K_{cmid} = 1.09\left[\frac{1 - \exp(-0.7(3.1))}{1 - \exp(-0.7(2.2))}\right] = 1.23$$

Therefore, the $K_{c\,mid}$ estimated for the complex of date palms and citrus together is 1.23. This value is compared with the maximum expected K_c based on energy limitations, represented by $K_{c\,max}$ of Equation which in this case for h = 10 m is $K_{c\,max}$ = 1.34. Because $K_{c\,mid} < K_{c\,max}$ (i.e., 1.23 < 1.34), the $K_{c\,mid}$ = 1.23 is accepted as the approximate estimate of the $K_{c\,mid}$ for the intercropped field.

Clothesline and Oasis Effects

Under the clothesline effect, where vegetation height is greater than that of the surroundings (different roughness conditions), or under the oasis effect, where vegetation has higher soil water availability than the surroundings (different moisture conditions), the peak K_c values may exceed the 1.20-1.40 limit.

The user should exercise caution when extrapolating ET measurements taken from these sorts of vegetation stands or plots to larger stands or regions as an overestimation of regional ET may occur.

Small expanses of tall vegetation that are surrounded by shorter cover can exhibit a clothesline effect.

This occurs where turbulent transport of sensible heat into the canopy and transport of vapour away from the canopy is increased by the 'broadsiding' of wind horizontally into the taller vegetation. In addition, the internal boundary layer above the vegetation may not be in equilibrium with the new surface.

Therefore, ET from the isolated expanses, on a per unit area basis, may be significantly greater than the corresponding ET_o computed for the grass reference. Examples of the clothesline or oasis effects would be ET from a single row of trees surrounded by short vegetation or surrounded by a dry non-cropped field, or ET from a narrow strip of cattails (a hydrophytic vegetation) along a stream channel. K_c values up to and exceeding two have been recorded for such situations.

Where ET estimates are needed for such small, isolated expanses of vegetation surrounded by shorter cover (clothesline effect) or dry land (oasis effect), then the K_c may exceed the grass reference by 100% or more. Estimates of K_c for the expanses of vegetation should contain u_2, RH_{min} and h parameters to adjust K_c values, and parameters expressing the aridity of the surrounding area, the general width of the vegetation stand and the ability of the wind to penetrate into the vegetation.

The equation should also consider the LAI of the vegetation to account for the ability of the vegetation to conduct and transpire the amount of water demanded by the clothesline/climatic condition. An upper limit of 2.5 is usually placed on K_c to represent an upper limit on the stomatal capacity of the vegetation to supply water vapour to the air stream under the clothesline or oasis conditions.

For vegetation with a great leaf resistance, such as for some types of desert vegetation or trees, the upper limit should be multiplied by a resistance correction factor, F_r.

Figure presents example curves of K_c for small areas of vegetation versus vegetation stand width, for conditions where u_2 = 2 m/s, RH_{min} = 30%, vegetation height = 2 m, and LAI = 3. The upper curve represents conditions where the specific vegetation is surrounded by dead vegetation, dry bare soil, or even gravel or asphalt. In this situation, large amounts of sensible heat are generated from the surrounding area due to the lack of ET. Some of this sensible heat is advected into the vegetation downwind.

The lower curve represents conditions where the vegetation is surrounded by well-watered grass. In this situation, there is much less sensible heat available from the surrounding area to increase ET (and K_c) of vegetation downwind. The influence of the aridity of the surroundings on the K_c for a small expanse is apparent. The two curves shown will change with changes in u_2, RH_{min}, h, and LAI. The user is cautioned that figure provides only general estimates of K_c under clothesline and oasis conditions. These estimates should be verified where possible using valid local measurements.

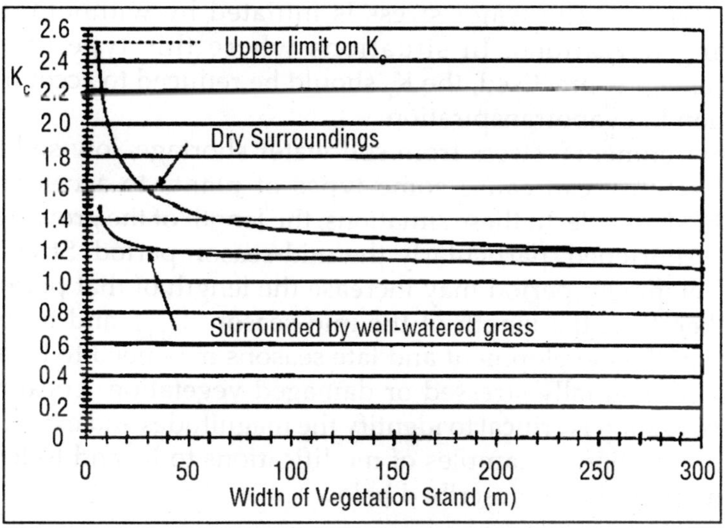

Fig. K_c Curves for Small Areas of Vegetation under the Oasis Effect as a Function of the Width of the Expanse of Vegetation for Conditions where RH_{min} = 30%, u_2 = 2 m/s, Vegetation Height (h) = 2 m and LAI = 3

ET estimates from large expanses of vegetation or from small expanses of vegetation that are surrounded by mixtures of other vegetation having similar roughness and moisture conditions should almost always be less than or equal to 1.4 ET_o, even under arid conditions.

For tall wind breaks, such as single rows of trees, an approximate estimate for K_c is:

$$K_c = \min\left(1.2 + \frac{F_r h_{canopy}}{\text{Width}}, 2.5\right)$$

where

F_r stomatal resistance correction factor

h_{canopy} mean vertical height of canopy area [m]

Width width (horizontal thickness) of the windbreak [m]

The K_c = 2.5 limit imposed in Equation 105 represents an approximate upper limit on ET_c of trees per unit ground area. However, this value has large uncertainty. Because availability of soil water may limit evapotranspiration from wind breaks, a soil water balance and calculation of the K_s stress factor should be conducted.

Management Induced Environmental Stress

Many agricultural crops are intentionally water stressed during specific crop growth periods to encourage particular crop characteristics. The water stress is initiated by withholding or by reducing irrigations. In situations where this type of cultural management is practised, the K_c should be reduced to account for the reduction in evapotranspiration.

Environmental stress from soil water shortage, low soil fertility, or soil salinity can cause some types of plants to accelerate their reproductive cycle. In these situations, the length of the growing season may be shortened, particularly the mid-season period. Stress during the development period may increase the length of that period.

Therefore, the length of the mid-season, L_{mid}, and perhaps the lengths of the development and late seasons may need to be adjusted for environmentally stressed or damaged vegetation. Local research and observation is critical to identify the magnitudes and extent of these adjustments. Some examples of modifications to K_c and to lengths of growing periods are described below.

Alfalfa Seed

Some forage crops such as alfalfa that are grown for seed production are intentionally water stressed to reduce the amount of vegetation and to encourage increased production of flowers and seed. In areas subject to freezing winters, the reduction in K_c for deep rooted crops such as alfalfa depends upon the amount of water made available from precipitation during the dormant (winter) season and upon the amount of rainfall and limited irrigation during the growing season. Therefore, the effects of the intentional stress on the values for K_c should be modelled using the basal crop procedure presented and the K_s coefficient and water balance procedure.

Cotton

In cotton production, soil water stress may be initiated during the development period to delay flower development and to encourage boll development. This practice retards the growth rate of the cotton plant and delays the date of full cover. For cotton, the attainment of full cover and the beginning of the mid-season generally occurs when the LAI reaches approximately three.

When soil water stress and growth retardation is practised, full cover may occur after the beginning of flowering. The effect of stress during the development period on ET_c can be incorporated into the K_c curve by extending the length of the development period into the mid-season period. The length of the total season generally remains the same.

Sugar Beets

Sugar beets are frequently managed to initiate mild soil water stress during the late season period to dehydrate roots and concentrate sugars.

A terminal irrigation may be needed just prior to harvest to assist in root extraction. When this type of water stress is practised, the value for $K_{c\ end}$ is reduced from 1.0 to 0.6.

Coffee

Coffee plants are often intentionally water stressed to reduce vegetation growth and to encourage development of berries. Under these conditions, K_c values from Table should be reduced. In addition, coffee fields may be bordered by trees that serve as windbreaks. The effect of windbreaks is to reduce the K_c of the coffee plants due to a reduction in wind and solar radiation over the plants.

The reduction in K_c could be significant where windbreaks are tall and frequent.

However, the K_c for the entire field area, including the windbreaks, may be increased by the presence of the trees, relative to the values for K_c for coffee shown in Table, due to increased total leaf area of the coffee-tree combination and the increased aerodynamic roughness.

Tea

Initiation and development of new leaves on tea plants often occurs following the start of the rainy season. During the dry season, initiation of new leaves is slow or non-existent. The transpiration from older leaves is lower than for new leaves due to effects of leaf age on stomatal conductance.

Therefore, the K_c, when leaves have aged (more than 2-3 months old), will be perhaps 10-20% lower than shown in Tables. Similar to coffee, tea fields may be bordered by trees that serve as windbreaks.

The effect of windbreaks is to reduce the K_c of the tea plants, but to potentially increase the K_c for the entire plantation, as described for coffee.

Olives

Growers may increase spacings of olive trees under rainfed conditions in areas with less rainfall. This is done to increase the ground area per tree that contributes infiltrated rain to transpiration of the tree. For example, in Tunisia, the spacing of olive trees changes from the north to the south, in proportion to annual rainfall. The tree spacing influences the K_c for the crop.

Chapter 10

Non-Growing Periods

This chapter describes procedures for predicting ET_c during non-growing periods. Non-growing periods are defined as periods during which no agricultural crop has been planted. In temperate climates, non-growing periods may include periods of frost and continuously frozen conditions.

TYPES OF SURFACE CONDITIONS

The type and condition of the ground surface during non-growing periods dictates the range for ET_c. Where the surface is bare soil, then the K_c will be quite similar to the $K_{c\,ini}$ predicted. Where the surface is covered by nearly dead vegetation or some type of organic mulch or crop residue, then the K_c will be similar to that for agriculture that uses a surface mulch.

Where the surface is covered by weed growth or growth of 'volunteer' plants, then the K_c will vary according to the leaf area or fraction of ground covered by the vegetation and by the availability of soil water. Where the surface is snow covered or frozen, then the K_c is difficult to predict and a constant value for ET_c may have to be assumed.

Bare Soil

Single Crop Coefficient

Where the ground is left mostly bare following harvest, then the K_c following harvest will be strongly influenced by the frequency and amount of precipitation. K_c for bare soil can be calculated as $K_c = K_{c\,ini}$.

Dual Crop Coefficient

Where a daily soil water balance can be applied, the user may elect to apply the dual K_c. In this situation, the topsoil layer may dry to very low water contents during periods having no precipitation. Therefore, the values for K_{cb} and for $K_{c\ min}$ in Equations should be set equal to zero.

This provides for the opportunity to predict $ET_c = 0$ during long periods having no rainfall. This is necessary to preserve the water balance of the evaporation layer and of the root zone in total. The daily water balance calculation, given $K_{cb} = 0$, will provide the most accurate estimates of ET_c during the non-growing periods.

SURFACE COVERED WITH DEAD VEGETATION

Single Crop Coefficient

Where the ground surface has a plant residue or other dead organic mulch cover, or where part of the unharvested crop remains suspended above the surface in a dead or senesced condition, then the surface will respond similarly to a surface covered by mulch.

In this case, K_c can be set equal to $K_{c\ ini}$ as predicted from Figure, but the value for $K_{c\ ini}$ can be reduced by about 5% for each 10% of soil surface that is effectively covered by an organic mulch.

Dual Crop Coefficient

Evaporation from dead, wet vegetation can be substantial for a few days following a precipitation event. Therefore, in the dual K_c approach, the value for f_c should be set equal to zero to reflect the lack of green cover and f_w should be set equal to 1.0 to reflect the wetting of both soil and mulch cover by precipitation.

The dead mulch or vegetation will dry more quickly than would the underlying soil if it were exposed. In addition, the soil will be protected somewhat from evaporation by the dead mulch or vegetation cover.

Therefore, total evaporation losses will be less than the TEW predicted from Equation. This can be accounted for by reducing the value for TEW by 5% for each 10% of soil surface that is effectively covered by an organic mulch. The value for REW should be limited to less than or equal to that for TEW.

Surface Covered with Live Vegetation

During frost-free periods following harvest, weeds may begin to germinate and grow. This vegetation is supplied with water from storage in the soil profile and from any rainfall. In addition, crop seed lost during harvest may germinate following rainfall events and add to the ground cover.

The amount of ground surface covered by vegetation will depend on the severity of weed infestation; the density of the volunteer crop; tile frequency and extent of soil tillage; the availability of soil water or rain, and any damage by frost.

The value for K_{cb} during the non-growing period can be predicted over time according to the amount of vegetation covering the surface. This can be done through estimates of LAI using Equation or estimates of the fraction of ground cover, f_c, using Equation.

Single Crop Coefficient

In the single crop coefficient approach, the value for K_{cb} determined using procedures can be converted into an equivalent K_c by adding 0.05 to 0.15 according to the frequency of rainfall and surface wetting.

It is important that the K_c for vegetation during the non-growing period be limited according to die amount of soil water available to supply evapotranspiration. Otherwise, the law of conservation of mass will be violated. Under all conditions, the integration of K_c ET_o over the course of the non-growing period cannot exceed the sum of the precipitation occurring during the period plus any residual soil water in the root zone following harvest that can be depleted by the subsequent vegetation.

The root zone in this case is the root zone for the weed or volunteer crops. A daily soil water balance may provide for the best estimate of soil water induced stress and associated reduction in K_c and ET_c.

Dual Crop Coefficient

Under the dual crop coefficient approach, K_{cb} can be predicted according to the amount of surface that is covered by vegetation. Then, a full daily soil water balance of the topsoil together with a full daily soil water balance of the root zone can be employed. The soil water balances will automatically adhere to the law of conservation of mass, so that total ET_c from the weed or volunteer vegetation will not be overestimated.

Again, because the topsoil layer may dry to below wilting point under conditions of sparse rainfall, the values for K_{cb} and $K_{c\,min}$ used in Equations should be set equal to zero. In this manner, the daily soil water balance with dual K_c calculations can progress throughout the non-growing period with good results.

Frozen or Snow Covered Surfaces

Where the ground surface is snow covered or frozen, any vegetation will be largely non-responsive and non-contributing to ET_c, and the amount of ET_c will be closely related to the availability of free water at the surface and to the albedo of the surface.

The albedo of snow covered surfaces can range from 0.40 for old, dirty snow cover to 0.90 for fresh, dry snow. Therefore, the ET_c for snow cover will be less than ET_o for grass, as 25-85% less shortwave energy is available. In addition, some energy must be used to melt the snow before evaporation.

The use of ET_o under such conditions is of limited value, as the assumption of conditions sustaining a green grass cover is violated. It is even possible to obtain negative values for ET_o on some winter days where the longwave radiation from the surface is large and the vapour pressure deficit is small. It is under these conditions that net condensation of water from the atmosphere is possible. This would be similar to negative evaporation.

Given the limited value of ET_o (or even ET_p) under snow covered or frozen conditions, a single, average value may be best used to predict ET_c. Wright (1993) found that ET_c averaged 1 mm/day over winter periods at Kimberly, Idaho, the United States, that were six months long (1 October to 30 March).

The latitude of Kimberly is 42°N and the elevation is about 1200 m. Over the six-year study period, the ground was 50% covered by snow for 25% of the time from 1 October to 30 March. The ground, when exposed, was frozen about 50% of the time.

The K_c averaged 0.25 during periods when the soil was not frozen but where frosts were occurring (October and early November). When the ground had 50% snow cover or greater, the ET_c averaged only 0.4 mm/day.

Wright found that over the six-month non-growing period, total cumulative ET_c exceeded precipitation by about 50 mm.

Figure shows the mean measurements of ET_c during the 1985-1991 study period. The measurements have high correspondence to the total shortwave radiation energy available on a clear day, R_{so}, estimated as 0.75 R_a.

There is some lag between ET_c and R_{so} and R_s caused by cooler temperatures in January - March as compared to the October - December period. The ET_c/R_{so} ratio averaged only 0.17 over the six-month period, and averaged 0.11 from 1 Dec. - 10 Mar. The ET_c/R_s ratio averaged 0.23 over the six-month period, and averaged 0.15 from 1 Dec. - 10 Mar.

A similar study conducted in Logan, Utah, the United States (latitude 41.6°N, elevation 1350 m) over an eight-year period showed that ET_c varied widely with soil surface wetness and air temperature during the winter months. The 'average' K_c from November to March was 0.5 for days having no snow cover. For days with snow cover, 'Et_c' ranged from 0 to 1.5 mm/day. Similarly, K_c is about 0.4 for winter wheat during frozen periods in the region of northern China (latitude near 39°N.

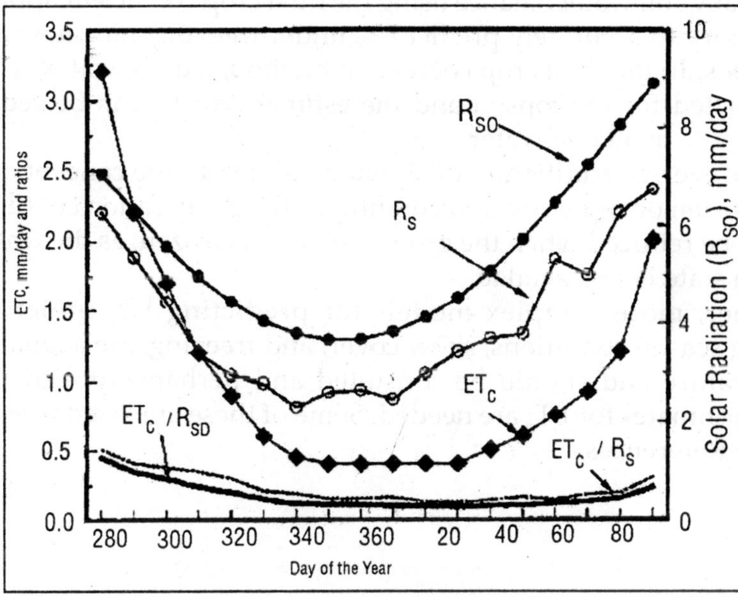

Fig. Mean Evapotranspiration Measured during Non-Growing

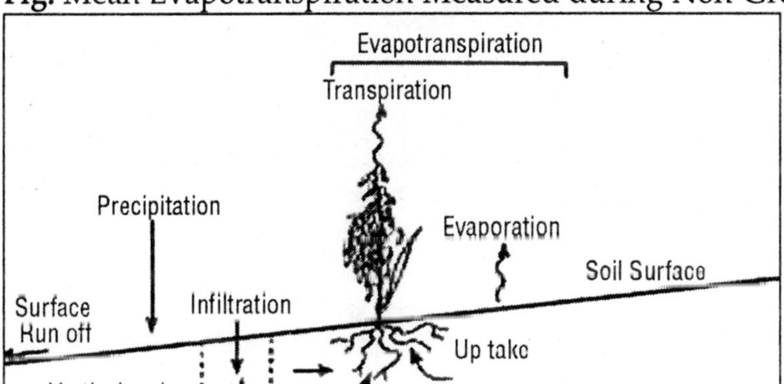

Single Crop Coefficient

The above procedure can provide estimates for the single K_c during non-growing season periods having snow cover or freezing conditions. However, the actual value for K_c is known to vary widely and will be less when water is less available at the soil surface.

Dual Crop Coefficient

A daily soil water balance using the dual crop coefficient approach is necessary to accurately predict ET_c under freezing and snow cover conditions. In the dual crop coefficient method, a daily water balance is conducted for the topsoil and the estimate for K_c can be reduced according to available water.

However, in addition to the limited validity of the concept of ET_o under frozen or snow covered conditions, the evaporation coefficient, K_e, may be reduced when the ground surface is frozen, as the water in a frozen state is less available.

Other, more complex models for predicting ET_c under non-growing season conditions, snow cover, and freezing, are available in the literature and should be consulted and perhaps applied when precise estimates for ET_c are needed. Some of these are listed in section K of the References.

Chapter 11

Environmental Impact on Crops

Soils not only respond to environmental conditions, but can also influence these conditions and therefore play an important role in the operation of environmental systems. These systems can be divided into four main types—the hydrosphere, atmosphere, geosphere and biosphere. Although these environmental components are in many ways interrelated, the role of soil in each system separately in order to provide a clearer understanding of the processes involved.

We will start by examining the hydrosphere because soil moisture is one of the main keys to soil-environmental influences, and many of the aspects relating to the hydrosphere are therefore relevant to those of the other systems. Having first introduced some basic concepts, various components of the hydrological cycle will then be considered.

Closely related to the hydrosphere is the atmosphere, with soils exerting an important influence on climate at or near the ground surface, and this is considered next. The geosphere is then examined with respect to the role of soil in geomorphic processes and landform development. Finally, we consider the biosphere, in which soils are an important factor in controlling the habitation and distribution of biota.

THE HYDROSPHERE

The activity of water at or below the ground surface is part of the hydrological cycle, in which water arrives at the surface via precipitation and is eventually returned to the atmosphere via evapotranspiration, having passed along a number of possible pathways in between.

Some precipitation may be prevented from reaching the surface directly by obstructions such as vegetation or buildings, and some may not reach the ground at all, being intercepted and returned directly to the atmosphere by evaporation. That water which does reach the surface will either remain there as *surface water,* flowing over the surface or being stored in depressions, or it will infiltrate the ground to become *subsurface water;* in this form it can be stored or moved through the ground, and may eventually become surface water.

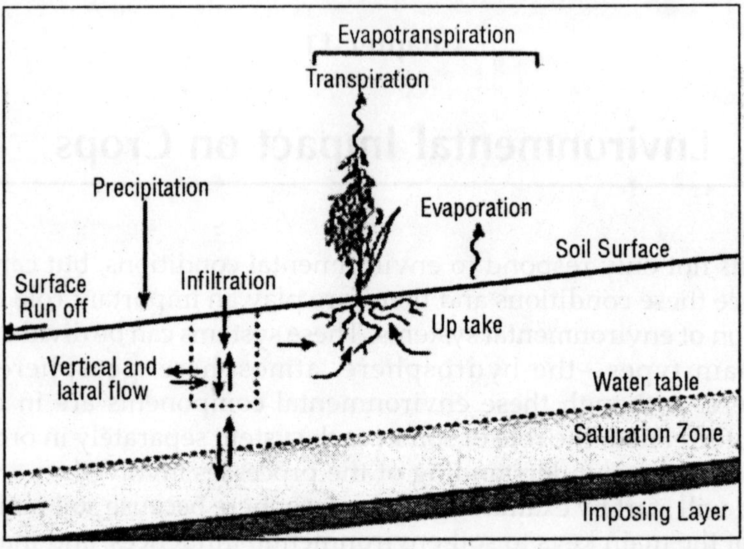

Fig. Components of the Hydrological Cycle in Relation to the Soil Layer.

Subsurface water can be classified into four major zones. The *soil zone* lies nearest the ground surface and therefore controls the infiltration of water into the ground. The underlying *intermediate zone* is one in which percolation of water is the dominant process and this can vary enormously in thickness depending on the relief and rock type.

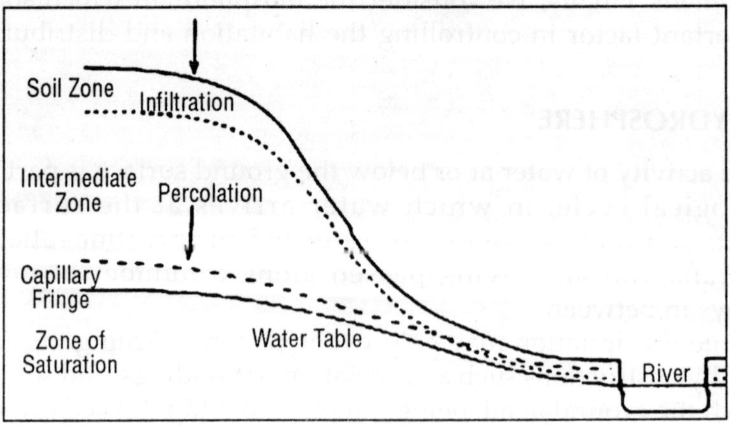

Fig. Classification of Subsurface Water

This overlies the *capillary fringe* in which most of the pores are filled with water, and beneath this lies the *saturation zone;* the interface of these two zones is marked by the *water table*. While these zones may be separated on interfluves and valley sides, they usually converge downslope and may overlap on valley floors.

In this section we will examine the influence of soils on the hydrosphere under three main headings—the forces controlling water movement, soil water additions and losses in terms of infiltration, percolation and evaporation, and water storage and flow. It is important to recognise that soils can also influence the chemical characteristics of water, but this will be considered in the context of the geosphere.

Forces Controlling Water Movement

The movement of water into, through and out of soils is controlled to a large extent by gravity, and also by three types of force determined by soil properties which can either encourage or restrict water movement—adsorption, capillarity and osmosis, the combined effects of adsorptive and capillary forces being known as *matric suction*. The total suction in a soil will depend to a large extent on its texture and pore size.

Higher clay or organic matter contents will have stronger adsorptive forces, and finer textures are also usually associated with smaller pore sizes, in which capillary forces will be greater. These factors therefore influence the retention and drainage of water. Coarse-textured soils generally drain more quickly and retain less water than finer-textured soils which, for any given water content, will have a greater suction.

Water content will, however, also influence suction. Dry soils will have high values, but these will rapidly decrease as the pores become filled with water and the suctional effects are therefore lost.

The speed and direction of water movement in a soil will depend on the magnitude of these suctional forces compared with the gravitational force. The rate at which water can move through a soil is measured by *hydraulic conductivity*, which is determined to a large extent by soil water content and also by texture and its associated pore size. Water transfer is more effective in wetter soils than in drier ones because drier soils have a greater volume of air in their pores which inhibits the conduction of water from one location to another.

Therefore at high moisture contents conductivity increases as texture coarsens; values can be less than 1 mm per day in the case of clays, 10 cm to 10 m per day in silty sand and over 100 m per day in gravel. However, at low moisture contents conductivity increases as texture becomes finer, because of higher suction forces.

The direction of movement under gravity is downwards, but soil water can move in other directions depending on the other forces present. For example, it can move laterally due to osmosis if there is a lateral variation in the concentration of salts dissolved in the soil water due to variations in parent material, or it can move upwards due to capillary forces operating as a soil dries out.

Infiltration, Percolation and Evaporation

When rainfall reaches the ground surface, some or all of the water will infiltrate the soil. Initially the matric suction gradient will be relatively high and infiltration will therefore be rapid, but as wetting increases the suction gradient lowers and the infiltration rate decreases as gravitational forces become more important and eventually a steady flow will be attained, approximating to the *saturated hydraulic conductivity*.

The *infiltration capacity* is a measure of the ease with which water can penetrate the surface. For example, because of their larger pores, coarse-textured soils will allow more rapid infiltration than finer materials. Infiltration will also be reduced if a crust has formed, for example by aggregates breaking down under raindrop impact, or if fines are washed into the surface pores.

The type and thickness of the litter layer can also affect infiltration; a thick cover of broadleaf litter with the leaves lying horizontally can reduce infiltration appreciably. Soil freezing will also impede infiltration as the pores become filled with ice, infiltration becoming zero in wet soils where the pores become completely ice-filled. Conversely, soils which develop cracks at the surface will have high infiltration rates, as in the case of Vertisols.

The percolation of water through the soil zone and intermediate zone towards the water table will continue while rainfall is infiltrating the surface, but once rainfall ceases the soil will start to dry out as water drains downwards and is also lost back to the atmosphere via evapotranspiration.

Drainage under gravity is usually more or less complete after two or three days and the soil is then said to have reached *field capacity*. The movement of water through a soil is enhanced by the presence of macropores; these are relatively large pores which can result from shrinkage of a soil on drying or from the development of burrowing or root channels.

They are particularly important in infiltration and percolation during intense rainfall, although the pores must be interconnected otherwise the continuity of flow is interrupted.

The loss of water from a soil by evaporation will be controlled by a number of factors, principally climate, moisture content and texture. High temperatures and windspeeds will enhance evaporation, as will high vapour pressure gradients between the soil and above-ground atmosphere.

The moisture content of the top few centimetres of soil is important in controlling evaporation, which will decrease as the soil dries out, becoming zero once the soil is completely dry.

In contrast the moisture content of the subsoil is considered to have little effect on evaporation because of the slow rate of soil moisture movement. Eventually a point is reached when little water is available for movement in the liquid form, and any remaining movement occurs by the diffusion of water vapour. At this stage the soil is said to be at *wilting point.*

Vapour diffusion is controlled by matric and osmotic pressure to a minor extent, but the most important control under most soil conditions is temperature, with vapour diffusion occurring from relatively warm to cooler areas of a soil in response to the vapour pressure gradient. However, the movement of water vapour from the subsoil to the surface constitutes only a minor proportion of total evaporative losses.

Hence, for any given volume of rainfall, soils which are regularly wetted at their surface will have greater evaporation values than those which are wetted more thoroughly but less frequently. Texture affects evaporation in that upward capillary movement of water is generally greater in fine-textured soils because of the greater suction forces.

In some cases water can move upwards through a vertical distance of several metres, although in coarse-textured soils the distance is unlikely to exceed several centimetres. However, because the speed of capillary water movement is slow, this source of water does not usually contribute greatly to total evaporation except where the water table lies within a metre of the surface.

Other factors which will affect evaporation include soil colour and vegetation cover. Dark soils will generally have higher surface temperatures due to their lower reflectivities and therefore experience higher rates of evaporation than lighter coloured soils. A vegetation cover can shade the soil surface and thus decrease surface temperatures and evaporation.

It can also increase the relative humidity of the air near the surface, which will lower evaporation. However, these reduced moisture losses may well be offset by the loss of water via transpiration from the vegetation itself.

Water Storage and Flow

Water can be stored in or on a soil, or can flow over its surface or within it. Storage of surface water requires a soil of low permeability and a surface relief which will prevent lateral drainage. The storage may be either temporary, as in the case of puddles formed in microtopographic depressions, or it may be permanent if the supply of water is sufficiently great to exceed losses via evaporation and infiltration; in such cases ponds or lakes may form.

The supply of water for surface storage can be provided directly by precipitation, water running downslope over the surface, or by ground water reaching the surface. Surface storage is therefore associated with fine-textured, compacted soils, and is favoured in cooler climates, where evaporative losses are limited. Water storage within the soil can occur where the regional water table approaches the surface, for example in low-lying areas or enclosed depressions.

Storage can also occur where local conditions cause waterlogging near the surface, as in the case where a clay-rich Bt horizon or an iron pan impede the downward movement of water; this is known as a *perched water table*. However, where soils are located on a slope, water flow will usually occur. Three main types of flow can be recognised—*overland flow, throughflow* and *ground-water flow*.

Overland flow occurs when the rate of infiltration is exceeded by the rate at which water is arriving at the surface. This can occur, for example, in the case of frozen soil conditions, or where aggregate breakdown and surface sealing are caused by raindrop impact where soils lack a protective vegetation cover.

Throughflow occurs when water flows laterally through the soil and is confined near the surface, as opposed to ground-water flow which involves water movement through the saturated zone beneath the water table, often at depths below the soil layer; this usually occurs more slowly than throughflow.

Throughflow can constitute a large proportion of total runoff from a catchment in cases where soils encourage infiltration and lateral water movement but restrict deeper percolation. Such conditions will occur, for example, if a soil has vertical cracks, extensive burrowing networks or other forms of macropore which allow rapid infiltration and also a humified organic layer, Bt horizon or some form of pan which impedes vertical drainage.

In extreme cases these conditions can lead to the development of piping, whose networks can provide major pathways for water movement through soils. Soil thickness and slope angle are also important in that thin soils over impermeable parent materials will encourage rapid throughflow, as will steep slopes. Even in the absence of such favourable factors for throughflow, soil hydraulic conductivity will usually be greater in the lateral than the vertical direction due to the development of structure and also to lower compaction than at depth, therefore on slopes, lateral flow will generally predominate over vertical drainage.

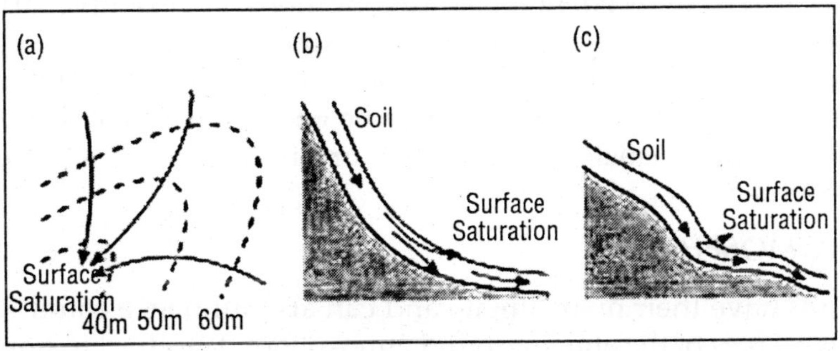

Fig. Principal Locations of Flow Convergence to Produce Surface Saturation: (a) Hillslope Hollow (Dotted Lines Represent Contours), (b) Footslope, (c) Soil thinning

Throughflow will be converted to surface runoff if the near-surface saturated zone reaches the surface. This can occur if the flow is concentrated in a hillslope hollow or towards the foot of a slope. Lateral thinning of the throughflow zone can also cause saturation to occur at the surface and therefore produce surface runoff.

The relative proportions of overland flow, throughflow and ground-water flow will determine the shape of the hydrograph of a catchment. A hydrograph is a plot of stream discharge against time, comprising one or more peaks, each of which has a rising limb and a recession limb. The rising limb represents an increase in discharge in response to an input of water to the catchment, usually by rainfall or snowmelt.

The discharge will reach a maximum value beyond which it gradually decreases through time, and this represents the recession limb of the hydrograph. The steepness and height of the rising limb of the hydrograph will therefore depend on the rate at which water can reach the stream channels. Where overland flow occurs, water will reach the channels quickly, whereas throughflow and ground-water flow will delay its passage.

The extent of the delay will be determined by slope angle and the way in which soil conditions influence vertical and lateral drainage. Therefore, soils on steep slopes and which impede infiltration will produce the most peaked hydrographs; soils which allow some infiltration but which have subsurface compaction, pans, Bt horizons or frozen subsurface layers, will tend to produce less peaked hydrographs, and soils which do not possess these features and therefore allow greater vertical infiltration will have the least steeply sloping and lowest hydrograph peaks.

It is important to recognise, however, that these trends can be complicated by antecedent moisture conditions. For example, dry soils will often produce a delayed discharge response compared to wet soils, regardless of their effects on the type of flow, although under very dry conditions a rapid response may be encouraged by hydrophobic coatings of soil particles, such as clay or organic coatings of mineral grains or peds.

THE ATMOSPHERE

Soils have their own climate and can also influence conditions in the lowest part of the above-ground atmosphere. The effect is, however, only on a small scale, and the study of soil-atmosphere interactions therefore forms part of the discipline known as *microclimatology*. The atmospheric components involved in this study will be considered under three headings—solar radiation and temperature, atmospheric moisture and air movement.

Solar Radiation and Temperature

Radiation reaches the ground surface from the sun, and leaves the ground surface, as electromagnetic energy of differing wavelengths. The wavelengths of the incoming radiation are predominantly shorter than those of the outgoing radiation, and these two types are therefore known respectively as *short-wave radiation* and *long-wave radiation*.

Some of the incoming short-wave radiation will be reflected off the ground surface back into the atmosphere; the amount reflected will depend on the reflectivity, or *albedo*, of the surface. The net all-wavelength radiation (Rn) can be shown as follows:

$$R_n = [S(1-a)] + L_n$$

Where S=incoming short-wave radiation, a=fractional albedo of the surface and Ln =net outgoing long-wave radiation. On unvegetated ground the albedo is determined by the nature of the soil. For example, dark soils, such as those rich in organic matter or derived from a dark-coloured parent material, can have an albedo of as little as 0.05 (or 5 per cent) and will therefore reflect little incoming radiation, whereas soils which are low in organic matter or derived from a light-coloured parent material can have values as high as 0.60 (60 per cent) and will therefore reflect much greater quantities of incoming radiation. Albedo also decreases with increasing water content due to internal reflection at the surfaces of water menisci in soil pores; a 20 per cent water content can decrease soil reflectivity to a quarter of its dry value. The angle of incidence of incoming radiation can also have a marked effect on reflectivity, particularly in the case of a soil with standing water at its surface, as will microrelief. If the angle of incoming radiation is low, a water surface will be much more reflective than for higher incidence angles; the albedo of water is only around 0.05 for incidence angles greater than 45°, but rises rapidly for decreasing angles.

Where soils have a marked microrelief, some of the incoming radiation will be absorbed by other parts of the surface, following the initial reflection, and hence the overall albedo of the surface will be reduced. The exchange of energy between the atmosphere and a surface can be shown by the surface energy budget equation:

$$R_n = G + H + LE$$

Where Rn =net all-wavelength radiation, G=ground heat flux, H= turbulent sensible heat flux to the atmosphere and LE=turbulent latent heat flux to the atmosphere. The *ground heat flux* represents the transfer of heat into the ground, while *turbulent sensible heat flux* to the atmosphere represents the transfer of heat by fluid (air) flow, and *latent heat* is heat absorbed during evaporation or released during condensation. During the day the available net radiation has a positive value and is balanced by turbulent fluxes of sensible and latent heat into the atmosphere and by conductive heat flux into the soil. Soils with a low albedo will attain higher temperatures at and close to their surfaces than those which reflect a higher proportion of incoming radiation, as seen in extreme cases for artificially whitened and blackened soils.

Higher surface temperatures will usually give higher above-ground temperatures in the first metre or so of air due to sensible heat transfer. Some of the heat will also be transferred by evaporation if the soil contains moisture, and some will be transferred deeper into the soil by conduction.

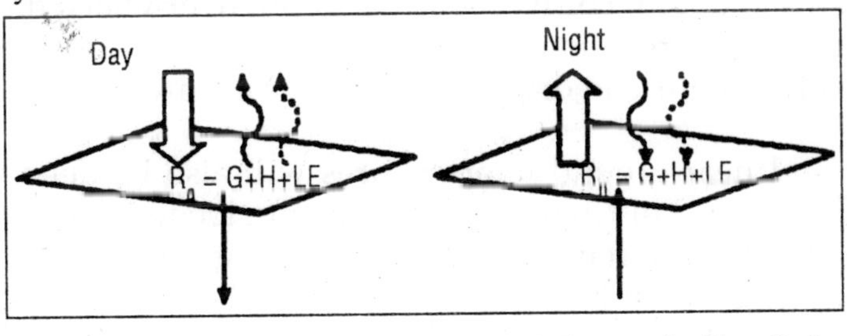

Fig. Energy Flows Involved in the Energy Balance of a Simple Surface

The rate at which heat transfer occurs through a material is expressed by its *thermal diffusivity (k)*. This is a function of its ability to conduct heat, known as its *thermal conductivity (k)*, and the amount of heat necessary to cause a temperature change, expressed by its *heat capacity* (C).

The relationship of these factors can be shown as: k =*k*/C, indicating that the rate of heat transfer is directly proportional to the ability to conduct heat, but is inversely proportional to the heat capacity. When saturated, a sandy soil therefore has a higher thermal diffusivity than a clay soil, which in turn has a higher value than a peat soil, but these values decrease when the materials are dry.

However, the greatest diffusivity will occur at intermediate moisture contents, because at high contents it is reduced due to the high heat capacity of water. Although a dry peat soil will have a low albedo because of its dark surface, little heat will therefore be transmitted deeper into it and the surface will become very hot. Sensible heat transfer will also cause the overlying air to heat up.

Conversely, a moist sandy soil is likely to have a higher albedo, reflecting more incoming radiation, and will also transmit more heat into the ground, therefore its surface temperature will be lower than that of the peat, as will that of the overlying air, but its subsurface temperature will be higher.

At night, the net radiation becomes negative due to the loss of outgoing long-wave radiation, and this is balanced by conductive heat supplied from the soil plus turbulent heat from the air. Under these conditions the surface of the peat will become colder than that of the sand because the peat is less able to replenish surface heat loss from below due to its lower diffusivity; this can cause a strong temperature inversion in the overlying air.

Most soils show greater diurnal and annual fluctuations in surface and overlying air temperatures than at depth, but soils with low diffusivity will have more extreme fluctuations in temperature at and near the surface than those with a higher diffusivity, and less extreme fluctuations at depth. For wet soils the depth over which diurnal and annual temperature fluctuations occur is around 0.5 m and 9.0 m respectively, whereas in dry soils the values are around 0.2 m and 3.0 m.

Soil and near-ground temperatures will also be determined by vegetation cover. This may have a different albedo from the soil and therefore reflect different amounts of incoming short-wave radiation. For example, grassland, heathland and scrub have albedos of around 0.15-0.25, while forests have values of around 0.05-0.20.

Vegetation also allows heat loss via transpiration in addition to evaporation. The effect of vegetation cover is to lower the diurnal range of soil surface temperatures; these are suppressed during the day, due to radiation reflection and heat absorption by the overlying vegetation, but enhanced at night due to long-wave emission and limited transpiration from the vegetation.

Similarly, the effect of soil on air temperatures is reduced by the presence of a vegetation cover. Snow cover will also affect soil radiation and temperature conditions. Because of its high albedo (0.95, decreasing to 0.40 with age), large amounts of incoming radiation will be reflected, and because of its low thermal diffusivity, similar to that of dry peat soil, heat exchange occurs at its surface with little heat being transferred to or from the soil.

Consequently, the soil is insulated by a snow cover from extreme temperature changes in the air, and the thicker the snow cover the greater this effect. Similarly, the effect of the soil on air temperatures will be suppressed under snow covers. Soil structure also influences thermal characteristics of the soil, with structured soils showing greater heat conduction than unstructured soil.

Atmospheric Moisture

Water occurring in a liquid form in the soil can influence the moisture of both the soil and above-ground atmosphere in terms of its water vapour characteristics. The amount of moisture which can be held as vapour in the air increases with temperature, and when the air is saturated with water vapour, condensation will therefore occur if cooling takes place.

The temperature at which this occurs is known as the *dew point*. Because air in pores of moist soils is in close contact with the water, it is usually close to saturation. However, vapour gradients will develop due to variations in soil temperature with depth. During the day soils are usually warmer towards the surface, therefore more water vapour can be held in the pores in the upper part of the soil than in those lower down. The resulting vapour concentration gradient produces a net flow of vapour down into the soil. Conversely, soils cool towards the surface at night which produces a net vapour flow upwards. If the vapour is cooled to its dew point on reaching the soil surface, condensation will occur; this process is known as *distillation*.

Moisture can also be added from the soil to the above-ground atmosphere via evaporation. This is greatest where temperatures are high, and is also enhanced by the desiccating effect of air moving across the soil surface, because fresh air is usually associated with lower humidities. Most evaporation occurs from the top few centimetres of soil, and once this water has been lost, further evaporation occurs by water being moved upwards by capillary suction, although this constitutes a relatively minor proportion of soil evaporative losses. The extent of upward capillary movement depends on pore size, small-pored fine-textured soils being more conducive to this process than larger-pored coarser-textured soils.

Moisture lost from the soil via evaporation will generally increase the vapour pressure of the above-ground atmosphere, the increase being greatest close to the surface and decreasing with increasing altitude. The increase will be greatest during the day when temperatures, and therefore evaporation, are at a maximum.

If at night the temperature of the soil surface falls to the dew point, condensation occurs on the surface; this process is known as *dewfall.* The exchange of moisture from the soil to the above-ground atmosphere will also be influenced by vegetation cover. In addition to direct evaporative losses, moisture will also be lost to the atmosphere via transpiration, and within the vegetation layer itself the reduction in turbulent transfer allows vapour to accumulate close to the soil surface.

Air Movement

Soils can influence the movement of air above the ground in two main ways. Air movement occurs when it is heated which causes it to rise, a process known as *free convection,* and variations in soil surface types will cause different amounts of heating and therefore different extents of thermal uplift. For example, low albedo soils usually have higher surface temperatures during the day than adjacent higher albedo types, and areas of dry soil attain higher surface temperatures than adjacent wetter areas because of differences in thermal diffusivity.

Because of such lateral variations in heating, the most favourable areas will experience greatest thermal uplift, which causes convectional circulation cells to develop, thus mixing the air and sometimes leading to the formation of cloud by condensation of the upward moving air. This mixing reduces vertical differences in temperature, humidity and wind speed, and if surface heating and convectional uplift are strong, the mixing effect allows faster-moving upper air layers to be brought down nearer the surface, thus increasing wind velocities close to the surface.

During the night a decrease in temperatures is accompanied by a decline in convectional activity, and therefore wind velocities near the surface also decrease.

The second way in which soil surfaces can affect air movement is by them causing horizontally moving air to be set in turbulent motion, a process known as *forced convection*. This is a smaller-scale process than the previous one, and the extent to which it occurs will depend on the aerodynamic roughness of the surface.

This can be expressed as *roughness length* and will be determined by factors such as the size and shape of aggregates if the soil surface is unvegetated, or by the type of vegetation cover. For example, a sandy desert soil surface has a roughness length of 0.0003 m, while soils more typically have values of 0.001-0.01 m; in contrast, short grassland has values of 0.003-0.01 m, while taller grassland has values of 0.04-0.10 m and forests have values of 1.0-6.0 m. Surface roughness will increase turbulence, although vegetation will reduce air movement within it; for example, forests can reduce wind speeds by over 90 per cent beneath the canopy.

Air movement within a soil occurs by *diffusion* whereby oxygen diffuses into the soil and carbon dioxide diffuses out. It can also occur by *mass flow,* in which gas molecules move from an area of higher to lower pressure. This can occur for a number of reasons. For example,

Turbulence can produce pressure differences which are equalised by air movements, some from above the ground and some from within the soil. Pressure differences can also occur within the soil due to diurnal temperature variations with depth. During daytime surface heating, warm air will rise from the upper part of the soil and this will be replaced by cooler air from below.

Conversely, cooling of the surface during the night will cause relatively dense air to sink into the soil. In contrast, diurnal variations in vapour concentration will cause downward movement of air during the day and upward movement at night. Plant roots can also extract water from pores, causing it to be replaced by air.

Air movement can be restricted where the topsoil is wet or frozen, causing increased levels of carbon dioxide. Slower gaseous exchange with the above-ground atmosphere at depth can also lead to higher carbon dioxide concentrations than near the surface. Carbon dioxide concentrations also vary seasonally, with greater quantities being evolved during higher biotic activity at higher temperatures.

THE GEOSPHERE

A variety of processes operate in the formation of a landscape, as shown in the hypothetical 9-unit landsurface model, in which a slope profile is divided into 9 units, each characterised by a set of pedological and geomorphological processes. The interfluve is dominated by vertical movement within the soil, while the upper slope shows both vertical and lateral soil movement.

The free face is characterised by weathering and rapid mass movement, and the midslope represents an area of downslope transportation, involving both mass movement and water flow. Unit 6 marks the zone of net deposition on the footslope while unit 7 is an area of deposition of material from both upslope and upvalley, via the river channel.

Units 8 and 9 are dominated by river channel processes operating down-valley. In this model, soils are therefore important in three principal ways—they influence the nature and intensity of weathering, surface and subsurface sediment and solute transport, and mass movement, and each of these aspects will now be examined in turn.

Weathering

Soils are an important factor in the supply of moisture which is required for many weathering processes. For example, thick soils, or those of a fine texture or on gentle slopes, generally hold more moisture and are often associated with greater extents of bedrock weathering. However, during its development a soil may reach a critical thickness beyond which little bedrock weathering occurs because the soil acts as a protective mantle rather than as a catalyst to weathering.

The variation in water availability can also be important in the case of weathering processes such as wetting and drying and salt crystal growth.

Soils with a fluctuating moisture regime are therefore likely to allow more effective weathering of the underlying bedrock in these cases than continuously dry or wet soils.

Soil thickness can also be important in terms of protecting the bedrock from fluctuating temperatures which can cause freeze-thaw weathering. Although climate will obviously play an important role in determining soil moisture and temperature characteristics, properties of the soil itself can also have a marked effect.

The chemical composition of water can influence its weathering effectiveness, and this can be determined by the soil through which the water has passed. Of particular importance in this respect is the pH of the soil and the speed at which water moves through it.

For example, water percolating through a soil of very high or low pH will generally have a greater weathering capability than that passing through a more neutral soil. Also, a fine-textured soil will hold the water for a longer period than a coarse-textured soil, and therefore allow a longer reaction time over which weathering can occur.

Sediment and Solute Transport

The transport of sediment as part of the denudation process can occur via the soil surface or subsurface, while solute transport tends to be confined within the soil. Surface sediment transport operates by aeolian processes, or by surface water movement; these processes have already been considered in the context of losses during soil formation, but soils can also influence the way in which these processes operate.

For example, soils with a high silt or fine sand content are most susceptible to wind erosion, which is also enhanced by low moisture contents and sparse vegetation covers as occur in arid and semi-arid regions. Soils themselves can also provide a source of material for aeolian abrasion of exposed bedrock.

The relative hardness of these materials will determine the effectiveness of abrasion, and since soils often contain many minerals with hardness values in the range 5-7 on the Mohs scale, these can be effective abrasives of relatively soft rock types such as shales and limestones.

Soils influence the erosion of sediment by water via their control on runoff. Low permeability soils will encourage rapid runoff, which can take the form of overland flow or throughflow. Overland flow is encouraged by soils with low permeability surfaces, particularly where vegetation cover is low so that interception of precipitation is limited and infiltration rates are therefore more likely to be exceeded by precipitation rates.

A limited vegetation cover will also expose aggregates at the soil surface to the direct impact of raindrops, which may cause disaggregation, resulting in blocking of the surface pores and therefore a decrease in permeability. Shallow soils will also encourage overland flow because of their limited water storage capacity. Unconcentrated flow or *sheetflow* will only occur on smooth surfaces, but on most surfaces their roughness causes the flow to be concentrated in the form of small ephemeral channels or *rills*, particularly on silty or clayey soils.

If these channels become sufficiently large to stabilise, gullies may develop which will then act as locations for concentrated erosion. Silt and fine sand will generally be most easily transported by surface wash, because clays and colloidal organic matter will resist movement due to their cohesiveness, while sand and gravel particles will be more difficult to move because of their greater weight.

However, in soils with a high sodium salt content, clays often become easily dispersed on wetting and drying, and this can encourage erosion and gullying, for example in southern Africa.

In the case of subsurface sediment and solute movement, the

hydraulic conductivity of a soil will exert an important control on the ease with which movement can occur. The potential for sediment movement will also depend on the size of material available for transport, the extent of its aggregation and the size of the soil pores, while solute movement will also be determined by the ease of dissolution of the soil components and the acidity of the drainage waters; in some cases, soils may inhibit solute loss by buffering the acidity, or by providing exchange sites which can remove dissolved ions from the drainage waters.

As in the case of surface runoff, material may be removed from a slope by subsurface processes in either an unconfined or a concentrated manner. Unconfined movement can occur throughout the soil as a whole, whereas concentrated movement operates in zones which develop in soils often in response to topographic conditions, forming patterns similar to those associated with surface runoff, although usually with more diffuse boundaries.

The smaller features are sometimes known as *percolines,* as distinct from the larger and more easily recognisable *seepage lines*. These zones can be important in the movement of solutes and fine particulate material. The removal of larger material can lead to the development of subsurface pipes, particularly in soils which are susceptible to vertical cracking, which allows water to penetrate rapidly into the subsoil; organic soils and those rich in expanding lattice clays are particularly prone to this condition.

Piping will also be encouraged in soils which have an abrupt decrease in porosity at some point below the surface, as this will inhibit downward water percolation and promote lateral subsurface flow.

Pipes can range from a few centimetres to several metres in diameter, and can therefore be responsible for transporting large quantities of water down hillslopes, along with sediments and solutes. If a pipe develops to a size which exceeds that capable of being supported by the overlying soil, the roof will collapse, exposing the channel to the surface. A similar situation may result from the extension of vertical cracking down to the level at which the pipes occur. These can be important mechanisms in the formation of gullies, especially in arid and semi-arid environments and also in some peat soils.

Soils can influence landforms by their resistance to sediment and solute transport. This is particularly the case in low latitude regions where soils have become indurated by the formation of pans. In these instances they can behave like resistant rock types and a number of geomorphological features can result from erosion.

For example, indurated layers, or *duricrusts,* can form cappings over softer materials, producing flat-topped interfluves, benches along valley sides or pavements on valley floors; they can also form escarpment features if inclined at an angle to the horizontal, or irregular relief due to differential erosion.

Studies of sediment and solute transport can be made at a variety of scales. For example, within a river catchment in mid-Wales, it was reported that physical denudation of the catchment was occurring at a rate of 3.5 mm per 1,000 years, based on measurements of river suspended sediment and bedload, while input-output budgets for solutes indicated that chemical denudation was operating at a rate of 2.9 mm per 1,000 years.

At a smaller scale, rates of sediment and solute transport can be measured by hillslope plot studies; for example, material eroded from a known area can be collected as it moves downslope, or erosion rates can be calculated from the redistribution patterns of radioisotopes within the soil.

Mass Movement

The influence of soils on mass movement depends largely on its mechanical and hydrological properties. Of particular importance are cohesion and friction, which can be described by the Mohr-Coulomb equation:

$$\tau = C + \sigma_{\Pi} \tan\phi$$

where *ô* = shear stress at failure, c = cohesion, *ó n* = normal stress on the shear plane and *ø*=angle of internal friction. The extent of cohesiveness and friction depends mainly on soil texture and moisture content.

In sandy soils there is little cohesion, so the shear stress at failure will depend on the normal stress and angle of internal friction. Conversely, in clayey soils cohesion will be high but friction low, therefore the shear stress at failure will depend on cohesion, while most soils have a combination of both friction and cohesion. However, moisture content will also affect stability because of the effect of the normal stress on pore water pressure. By taking this into consideration, the previous equation can be rewritten as:

$$\tau = C + (\sigma_{\Pi} - u) \tan\phi$$

or:

$$\tau = C' + \sigma'_{\Pi} \tan\phi'$$

Where *u* = pore water pressure, and c^2 and *ø′* are the cohesion and friction angle with respect to the effective stress. If pore water is therefore compressed by a soil on a slope, it will come under pressure and the value of will decrease, thus reducing the shear strength. Conversely, negative pore water pressure, or suction, can increase soil strength; for example, damp sandy soil in which capillary forces are acting will be more stable than dry sandy soil in which there will be no such effect.

Mass movement comprises three main sets of processes—heave, flow and slide—and various types of movement can be identified according to the relative combination of these processes. Transfer by rivers, rockslide and talus creep are not strictly relevant in the context of soil influences, but soils can have important effects on the remaining processes.

Soil creep can occur in one of two forms—*continuous creep* and *seasonal creep*. Continuous creep is mainly confined to materials with low shear stress values such as clay-rich soils. Movement is usually most rapid near the surface and decreases progressively with depth. Experts report movement to depths of 10 m and rates of movement up to 22 cm a^{-1} in the upper layers. Seasonal creep is more common and results from forces which cause variations in shear stress such as expansion and contraction caused by wetting and drying or freeze-thaw, plus disturbances due to bioturbation.

Movement is often confined to the top metre of soil in which these processes are most marked, and can occur in any direction. Rates of downslope movement are therefore difficult to establish, although values in the order of 0.5-2.0 mm a^{-1} have been suggested for humid temperate regions and up to 15 mm a^{-1} for temperate continental climates.

Soil creep can produce small step-like features known as *terracettes,* which run roughly parallel with the contour. The origin of these features has been the subject of some debate, and suggested alternative mechanisms to soil creep have included micro-scale slumping and trampling by animals.

Gelifluction (solifluction) occurs by movement of material over a frozen subsoil. When ice lenses in the upper part of a soil melt, pore water pressures are increased if the water is unable to drain away because of the frozen subsoil, and this can reduce the shear stress at failure of the material, causing it to become unstable and move downslope.

Expansion and contraction by freeze-thaw may form an additional component of the movement, as in the case of soil creep, as may the process of *frost creep*. The latter has been suggested to occur when particles on a sloping soil surface are displaced upwards at right angles to the slope when needle ice forms, and then on melting are returned to the surface, under gravity, via a more vertical path. Repeated freezing and thawing therefore produces a zig-zag movement of particles down the slope in the vertical plane.

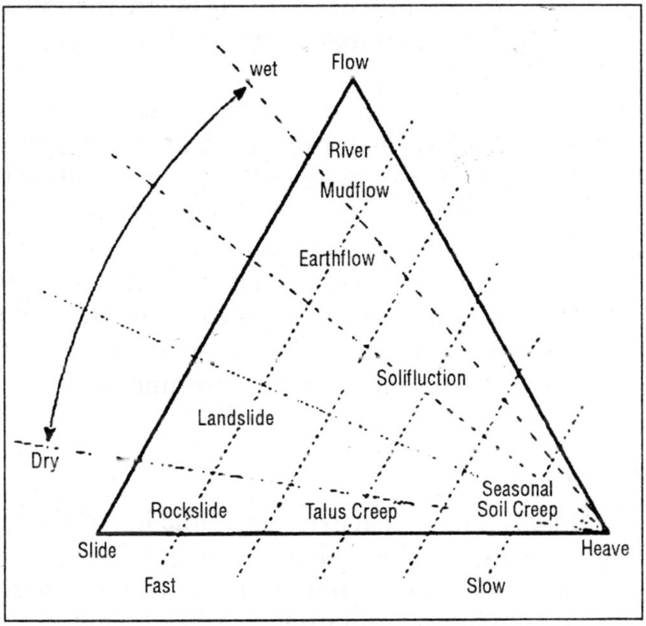

Fig. Classification of Mass Movement Processes

However, the use of the term *creep* to describe this process has been criticised on the grounds that it has nothing to do with deformation of material in relation to shear stress, and its operation within a soil as a whole, as opposed to simply the surface particles, has also been questioned. Silty soils are often most susceptible to gelifluction as these are prone to large quantities of segregation ice development so that large volumes of water will be released into the soil on melting.

Gelifluction material can occur as a featureless sheet if moving at a similar rate over an entire slope, or as terraces or lobes if moving at different rates or confined to particular parts of a slope. In the latter case gelifluction may be confined to areas downslope of late-lying snow patches which can increase the water content of the soil and therefore contribute to its instability.

Rates of gelifluction vary widely, but are generally greater than those of seasonal soil creep, being typically in the order of 0.5-10 cm a^{-1}, while depths of operation will depend on the extent of thawing.

Landslides, earthflows and mudflows are generally more rapid movements than gelifluction and soil creep, and usually result from instability caused by the reduction of shear stress at failure by high pore water pressures. They are therefore often associated with clay-rich soils with low angles of internal friction, or with soils which receive sudden influxes of large volumes of water via heavy rainfall; if removal of the water is impeded by an impermeable bedrock layer beneath the soil this will enhance instability.

Soils with pans or other forms of low-permeability illuvial horizon can also act in a similar way, as for example in the case of podzol Bs horizons. Rates of rapid mass movement vary enormously from 1 cm day^{-1} to over 1 m s^{-1}. Well-developed root networks in soils can increase the stability of slopes which are prone to slippage.

In addition to the forms, mechanical transfer processes operating within the soil itself can produce small-scale microrelief and patterned features. For example, frost-susceptible soils can produce earth hummocks and sorted patterns in cold environments, while Vertisols can also produce mounds and hollows.

THE BIOSPHERE

Soils are influenced to a great extent by biota in terms of pedogenic processes. However, this is by no means a one-way relationship—soils can have an important effect on biota in terms of the extent to which they encourage or constrain habitation.

The principal influences of soil on biota are via nutrients, moisture and aeration, although a number of other controls, both chemical and physical, are also important. For ease of discussion these factors will be examined individually, although it should of course be recognised that in reality many of the factors operate in combination.

Nutrients

The nutrients required for plant growth fall into two types—the *macronutrients,* which are used in relatively large quantities, and *micronutrients* or *trace elements,* used in very small amounts. Plant growth can be retarded if these occur in insufficient quantities, are not balanced by other nutrients or do not become available sufficiently quickly. Micronutrients are required by plants in only small amounts, but they are no less important than macronutrients.

While many nutrients are derived from the weathered mineral soil, organic matter can also be an important source, particularly in the case of nitrogen, phosphorus and sulphur, nutrients which are often in short supply. Nutrient content will therefore depend on the type of mineral and organic components which make up a soil.

For example, potassium, calcium and magnesium will be plentiful in soils rich in feldspars, mica and hornblende, while nitrogen will occur in higher quantities in organic material rich in proteins and amino acids. Sulphur can occur in minerals such as pyrite and gypsum, in addition to organic matter, and phosphorus can occur in the mineral apatite as well as in nucleic acid and other organic forms.

Consequently, nutrient-rich soils are generally associated with base-rich parent materials such as limestone, marl, basalt and some sandstones, and also with grassland or broadleaf trees; soils developed from parent materials such as granite, gneiss and quartz-rich sandstones, or those supporting heathland or needleleaf trees are usually nutrient-deficient.

However, nutrient availability can be equally important as nutrient content. For example, nutrients may be easily lost by leaching in a coarse textured soil, even if developed from a nutrient-rich parent material. Nutrient availability is also determined by the quantity held in an exchangeable form on clay and organic colloids, from which they can be removed by roots.

For example, calcium is normally held in this form in much larger quantities than potassium and magnesium, but consequently it is also more prone to loss by leaching, especially in freely draining soils. In the case of nitrogen, phosphorus and sulphur, their availability will also depend not only on the type of organic matter but on its rate of decomposition; if this is slow, due for example to low temperatures or poor drainage, these elements may not be released in sufficient quantities for certain plants.

Nutrient deficiency can be manifested in plants in four main ways. These are *chlorosis* (yellowing of leaves due to a lack of chlorophyll), *purpling* (darkening of leaves due to the accumulation of anthocyanin pigments), *local necrosis* (death of tissue) and *stunting* (reduced growth).

Chlorosis can result, for example, from a deficiency in nitrogen, phosphorus, magnesium, iron, zinc or copper, while purpling is often associated with phosphorus deficiency. Local necrosis can result from potassium or molybdenum deficiency, and stunting of leaf growth can occur due to zinc deficiency.

Nutrients also exert an important influence on soil mesofauna and micro-organisms. In terms of mesofauna, the principal source of nutrients is organic matter occurring as either living or dead plant or animal material. In the case of herbivorous fauna, the nutrient status of the plant-derived organic matter will determine population levels, with higher levels usually present in soils with high nitrogen content organic matter, such as those associated with many grassland and broadleaf tree types.

Carnivorous mesofauna will occur in greatest quantities in soils which have large populations of animals on which they prey, and these are therefore usually the same types of soil. For soil micro-organisms, carbon and nitrogen are the nutrients which are most commonly deficient, and various adaptations have been developed in order to overcome this problem; these include the fixation of atmospheric nitrogen, the ability to survive long periods without nutrition, and the release of antibiotics which inhibit the growth of competitors. Another adaptation is autotrophy, or by the oxidation of ammonia, sulphur or sulphide.

Moisture and Aeration

Soil moisture is required for the growth of most plants, and the moisture content of a soil is determined to a large extent by drainage conditions. For example, a coarse-textured soil, or one with a well-developed open structure, will usually drain more easily than a finer-textured or compacted soil because of its greater hydraulic conductivity, and will therefore be more prone to drought.

However, clay-rich soils such as Vertisols, which contain expanding lattice minerals, may also have low moisture contents because of large vertical shrinkage cracks which form during periods of dry weather, causing rapid infiltration and percolation through the upper part of the profile. Similarly, shallow soils or those on steep slopes often have low moisture contents. In arid environments, slope position is particularly important, with footslope soils usually providing the best water supply.

The availability of moisture to plants is, however, determined not only by the moisture content of a soil, but also by the suction at which the water is held. For a given moisture content, the forces holding water in the soil, which must be overcome if a plant is to obtain water, are lowest for coarse-textured soils which have relatively large pores, and greatest for fine-textured soils in which the matric suction effects of the small pores are much higher.

For any given texture, the forces also decrease as moisture content increases; sand has the least water available to plants (only 7 per cent) while silt loam has the most (16 per cent). In order to survive in moisture-deficient soils, plants have developed a variety of adaptations, for example deep and closely spaced roots to maximise water acquisition, and narrow, waxy or hairy leaves to minimise water loss from transpiration.

In addition to plant uptake, moisture is required to allow penetration of roots and burrowing organisms; when soils are dry they often become hard and compact, thus restricting these activities. In contrast to plants and mesofauna, micro-organisms do not generally experience problems in dry soils, and indeed some types, especially spore-forming bacteria, can withstand drought conditions for many years, although bacterial growth rates usually increase when dry soils become moist.

Excessive moisture can cause different, but equally severe problems for biota. Soils can be prone to excessive wetness for a variety of reasons; for example, heavy rainfall, low permeability or poor drainage due to topographic location. Under such conditions, most pores are filled with water and therefore the root zone becomes anaerobic, which can have serious effects on the metabolism and growth of plants not adapted to such conditions, even if the conditions last for only a few days.

Root systems of these plants require oxygen partial pressure of at least 0.02 bar, with nutrient uptake increasing as partial pressure increases, up to a value of around 0.2 bar. When root cells become affected by a lack of oxygen, they start to convert glucose into ethanol (C_2H_5OH), causing cell membranes to become prone to leakage, and ion and water uptake to be impaired. Consequently wilting can occur, accompanied by chlorosis.

Also under waterlogged conditions, anaerobic bacteria produce ethylene (C_2H_4), which can inhibit root growth at concentrations of 1 ppm or more, and volatile fatty acids can also be produced during carbohydrate decomposition, which can accumulate to harmful concentrations.

Anaerobic micro-organisms can cause denitrification, reducing nitrate to nitrites, or in more extreme cases to nitrous oxide and nitrogen, therefore causing the loss of this important nutrient. An additional problem with this reaction is that an accumulation of nitrite nitrogen can be toxic to plants.

In poorly drained gley soils, manganese and iron become reduced, in which form they are readily soluble and can be taken up by plants in toxic concentrations. Soils which experience periodic or prolonged flooding can pose an additional problem to plants, but those adapted to such conditions are able to transfer oxygen rapidly to their roots via spongy stem tissue, thus maintaining an aerated rhizosphere.

Soil micro-organisms are also inhibited by a restricted oxygen flow, as can occur in clayey soils, although certain groups are adapted to anaerobic conditions, but their growth is usually slower than those found in aerobic soils. Soil meso- and macrofauna, however, experience problems in extremely wet soils due to a lack of oxygen and instability of burrowing passages; in such soils the fauna are therefore concentrated near the surface, with few, if any, burrowing animals.

Additional Factors

Although nutrients, moisture and aeration are important influences on most biota, a number of other factors, both chemical and physical, can exert an influence on these properties and also act as additional influences in their own right.

The chemical factors relate principally to soil pH and toxins, while the physical factors include temperature, soil depth and ground stability, and soil pore size, compaction and induration. Soil pH can affect biota both directly and indirectly via its influence on nutrient availability and toxicity.

The direct influence is often not great because most plants can tolerate a wide range of pH values as long as nutrients occur in sufficient quantity, although at very low pH, hydrogen ions become toxic to plants. However, the effect of pH on nutrient availability and toxicity can be major. For example, iron, manganese and zinc are less readily available at alkaline than acid pH values, while calcium and molybdenum availability is greater at higher pH values.

Phosphorus is often in short supply because it occurs in forms which are not readily soluble, but it is most easily extracted by plants at around pH 6.5. At pH values below about 5.0, aluminium, iron and manganese can become soluble to an extent which can make them toxic to certain plants.

For example, aluminium toxicity inhibits cell division in root tips, causing them to become stunted, while manganese toxicity produces leaf distortion and yellowing. Low pH is also associated with calcium deficiency, which can cause leaves to wilt and collapse. Conversely, at very high pH values, bicarbonate ions can occur in concentrations which reduce nutrient uptake. Carbonates can also accumulate as precipitates around roots, which inhibits water and nutrient uptake.

The type of organisms inhabiting a soil can also be influenced by pH. For example, earthworms, bacteria and actinomycetes prefer neutral to slightly alkaline soil conditions, whereas fungi generally require more acid conditions. There are exceptions, however; iron- and sulphur-oxidising bacteria can occur in extremely acid conditions, with pH values as low as 1 or 2.

Some substances can also occur at toxic levels independently of pH conditions. Perhaps the most important of these are sodium salts, which increase the osmotic gradient that plant roots have to overcome in order to absorb water. If this gradient cannot be overcome, *plasmolysis* will occur in which movement of water out of plant cells towards the salt solution will cause the cells to collapse.

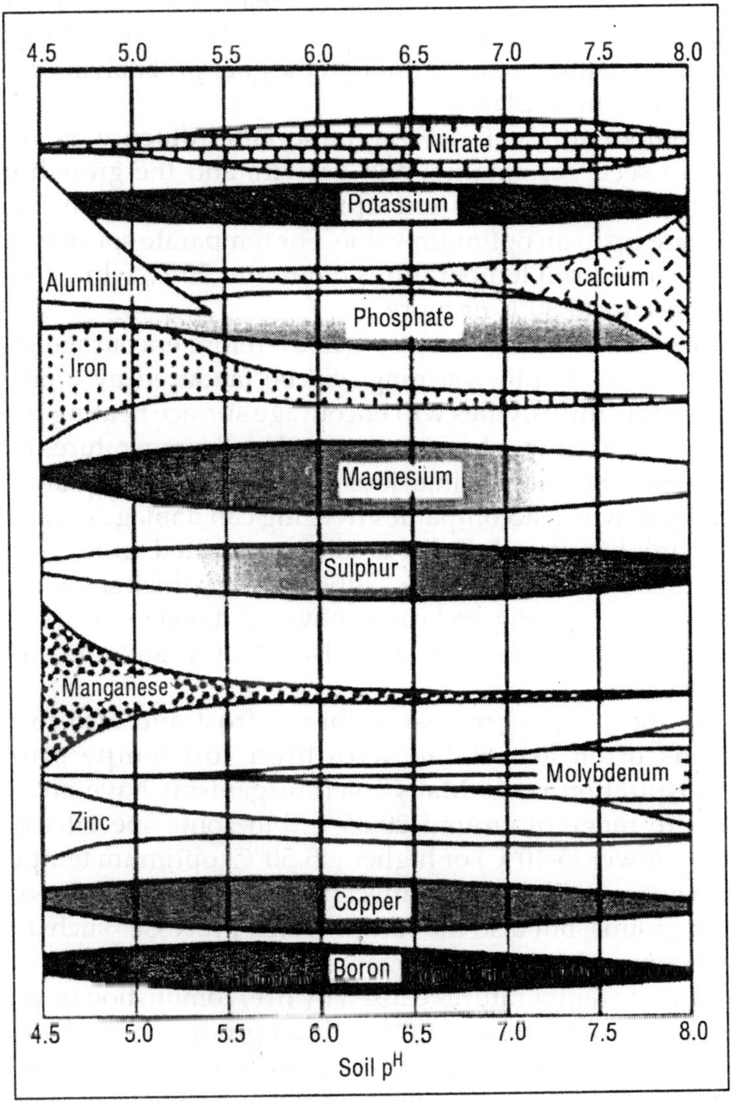

Fig. Relationship Between Soil pH and Plant Nutrient Availability in Mineral Soils

Soils with high sodium concentrations, such as those occurring in arid or coastal areas, therefore support plant communities which are specially adapted to high osmotic pressures. Other naturally occurring toxins include copper, chromium, arsenic and mercury, although these are normally found at toxic concentrations only in soils developed from metalliferous parent materials, or close to sources of industrial contamination.

Toxins derived from vegetation can lead to lower populations of soil fauna, as in the case of earthworms. Micro-organisms are also susceptible to natural toxins, although they are probably more tolerant of salinity than most plants.

Soil temperature will have an important influence on vegetation in terms of seed germination and growth, and the growth of roots, with both water and nutrient uptake increasing with increasing temperature up to an optimum value. For temperate species, optimum growth occurs at soil temperatures of around 20°C, while for tropical species it occurs at 30°C or more.

Soils which warm up rapidly will therefore encourage growth to commence early. Rapid warming will be favoured by high thermal diffusivity; low diffusivities will encourage surface heating but inhibit heat penetration of the subsurface layers. If soil temperatures fall below freezing point this may be injurious to certain types of plant, because the expansion which accompanies freezing can damage organic tissue.

This problem is reduced if a soil is protected from sub-zero air temperatures by a cover of snow, although very thick snow covers can both delay soil warming and cause physical damage to vegetation by compression. Soil temperature can also affect mesofauna and micro-organisms.

Many mesofauna are susceptible to frost and also to drought conditions often associated with high soil temperatures and evapotranspiration rates. Many micro-organisms have an optimum temperature range of around 20-35°C, but some specialised groups have much lower (5-10°C) or higher (45-50°C) optimum temperatures. Soil temperature will therefore affect not only the populations of micro-organism groups, but also the rate at which processes such as organic matter decomposition and nitrogen fixation occur.

At extreme temperatures caused by fire, combustion of vegetation and organic accumulations in the soil can produce a marked increase in nutrient content, and therefore in soil fertility, although this can quickly decline due to subsequent rapid leaching and erosion.

Shallow soils or those on steep slopes can limit moisture availability, but can also limit vegetation growth in extreme cases due to restricted root penetration or instability of the ground. Plants adapted to such conditions are therefore usually not only drought-resistant but also shallow-rooted and fast-growing.

However, shallow soils do not always limit root depth; in some cases roots can penetrate great distances into the underlying bedrock.

Shallow soils or those on steep slopes may also discourage habitation by burrowing macrofauna because of the potential instability of burrowing passages. Instability on more gentle surfaces can also limit plant growth, as in the case of the removal of topsoil by wind or water erosion, or mechanical disturbances such as cryoturbation.

Soil pore size and interconnectivity, and soil compaction and induration, in addition to their effects on soil moisture and aeration, will influence plant growth in terms of the ease of root penetration through the soil. Many roots exceed 60 ìm in diameter, but in fine-textured soils many of the pores are much smaller than this, therefore these will be impenetrable; root penetration is therefore confined to the larger spaces between individual particles and aggregates.

Soils with compacted or indurated layers will similarly restrict root extension. This can be seen for example in the case of iron pans, which can cause shallow tree root systems to develop above them, leading to instability in strong winds.

Chapter 12

Soil Factors in Crop Production

Landuse systems are open systems characterised by inputs, outputs, transfers and transformations of material and energy, and operating in parallel with hydrological, nutrient and energy cycles. In a general context, landuse systems may be classified as agricultural or non-agricultural. Agricultural systems may be further subdivided into arable, pastoral and forestry practices, while non-agricultural systems focus on urban, industrial, recreational and leisure activities.

The diversity of landuse practices reflects the influence of a complex array of physical and human factors, and soil occupies a central position within this array.

The choice of land use in a given location is therefore based largely on the extent to which soil characteristics match landuse requirements. Other important physical ,controls include climate and relief, while important human controls include socio-economic status of the land user, the nature of

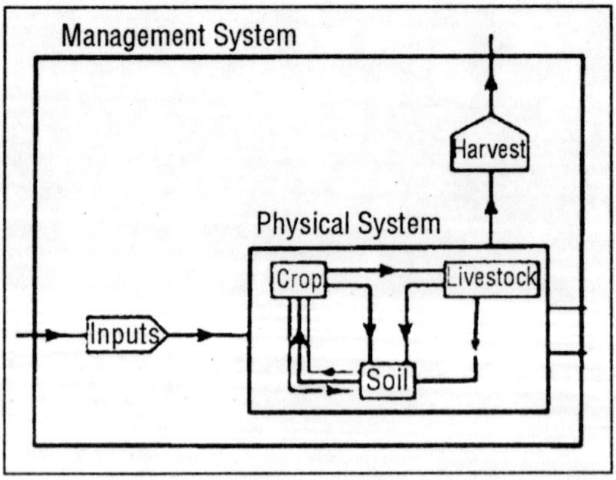

Fig. Generalised Structure of an Agricultural System and its Relationship with External Environmental Systems Land Tenure, and the Political and Infrastructural Characteristics of the Country and Region Concerned.

Landuse practices vary dramatically not only in their operational

details, but also in the scale of their operation. Individual activities may vary from a few hectares in area, as in the case of small forest clearances for shifting cultivation, to several square kilometres for mechanised cereal farms, and several hundred square kilometres for grazed rangelands.

In addition, landuse activities are often integrated in various ways in an attempt to increase their flexibility and long-term sustainability. Examples include strip grazing in fields with arable crops, combinations of trees with arable crops (agroforestry) and integrated arable, pastoral and forestry activities managed at the watershed scale.

Land not developed for productive agricultural purposes, such as mountain, moorland, wetland and other wilderness areas, may be used for recreational activities, or it may be protected for conservation. Non-agricultural land may also be used for urban and industrial development. In some circumstances, pressure on land is so great that reclamation of previously unused land is necessary.

This has been particularly widespread in low-lying coastal, estuarine and floodplain areas in heavily populated regions, including the Netherlands, and valleys of the Nile, Indus and Mississippi rivers. Similarly, there is increasing pressure in many areas to restore to its former quality, disused and derelict land, which has been disrupted by activities such as opencast coal mining, mineral extraction, waste disposal and construction.

In addition to their inherent spatial diversity and complexity, landuse systems are the product of numerous changes and developments which have taken place over a variety of temporal scales, more particularly during the last half century. In relation to arable systems, these changes include the evolution of advanced crop rotation systems, increased mechanisation, development of new crop varieties, improved agrochemical treatments, and the development of more effective drainage and irrigation technology.

More recently, energy- and cost-saving developments include reduced and zero-tillage practices, and low external input systems. In pastoral systems, new breeds of livestock have been developed, together with improved varieties of grass, and more effective pasture improvement strategies.

In relation to forestry, highly productive plantations of rapid growing species have been established in many areas, and in recent decades agroforestry and social forestry programmes have also been developed. Inevitably, all of these changes and developments have important implications for the maintenance of soil quality.

Clearly, the topic of soil-landuse relationships is enormous in its scope. It can be examined over a range of spatial scales, and is of interest to a variety of different disciplines. For example, an agronomist may wish to study the performance of crop plants in different growth media using pots or trays in a glasshouse experiment.

Similarly, a soil physicist may be interested in the impact of different tillage regimes on the physical characteristics of soils; this could be tested at the farm scale using clearly identified plots of land which are similar in all respects apart from their tillage treatments. At the larger scale, an environmental manager may be concerned with sediment yield or soil acidity variations between catchments under different land uses, while a geographer may be interested in the spatial relationship between soils and land use at the regional or national scale where broad soil types, rather than specific soil properties, may be important.

The factors which influence landuse systems will be examined, with particular emphasis on the role of soils. A variety of examples of practices within each of the major landuse systems—arable, pastoral, forestry, urban/industrial and recreation/leisure—will then be considered in terms of their effects on the soil. Emphasis will be placed on broad soil-landuse relationships rather than on details of the processes involved. For a more comprehensive discussion of this topic more specialist texts should be consulted.

INFLUENCE OF SOILS ON LAND USE

Soil plays a central role in the effective operation of landuse systems. In agricultural systems it has an influence on the yield of produce from arable, pastoral and forested land, particularly through its control over conditions within the root zone (rhizosphere). It also affects the nature and timing of mechanical operations and animal stocking levels.

In addition, soil has an influence on the behaviour of fertilisers and pesticides, and other related agricultural chemicals; consequently the susceptibility of certain crops to pests and diseases is influenced by soil characteristics. In non-agricultural landuse systems, the role of soil is just as important as in agricultural systems.

For example, in the engineering and construction industries, the success of mechanical operations, and the stability of resulting structures, depends on the mechanical characteristics of soils and their parent materials. Soil characteristics also have an important bearing on waste disposal practices, and on land reclamation and restoration procedures.

Many of the soil characteristics which influence landuse practices are interrelated to a greater or lesser extent. Hence, statistical associations between individual soil properties and landuse parameters, such as crop yield, are often unclear. Nevertheless, it is possible to examine some of the broader influences of the more important soil characteristics on landuse potential.

Soil Physical Factors

Texture is one of the most stable soil properties and has a major influence on landuse capability. Clay soils display relatively good water and nutrient retention but are often characterised by poor drainage and aeration. Furthermore, because such soils are often wet, they respond only slowly to changes in air temperature. Consequently, in clay soils of temperate environments, crop germination and emergence may be delayed in spring, although the onset of autumn frost is also delayed.

Clay soils are frequently difficult to cultivate and are thus said to have a 'heavy' texture. Workability and trafficability are particularly restricted during the autumn and winter periods. As clay soils are characterised by the development of large aggregates, however, they tend to be relatively resistant to erosion.

In contrast to clay soils, sandy soils display relatively poor retention of water and nutrients, and tend to be freely or excessively drained, and well aerated. Consequently, they respond rapidly to changes in air temperature. Sandy soils are generally easy to cultivate and they are said to have a 'light' texture. Aggregation is poor in such soils, however, and they may be susceptible to erosion, particularly if they are dominated by fine sand and contain little organic material.

Similarly, silty soils are prone to crusting and capping, which results in reduced infiltration, increased frequency of overland flow and increased risk of erosion. In an agricultural context, medium-textured soils, especially the finer silt and clay loams, are the most desirable. Such soils display optimum conditions in terms of water and nutrient retention, drainage and aeration, response to changes in air temperature, resistance to erosion and ease of cultivation.

In terms of landuse capability, soil structure is one of the most important soil characteristics, and is best developed in soils which contain appreciable quantities of clay and humus. Aggregates provide the basis for nutrient and water retention, while inter-aggregate pore networks facilitate aeration, rapid drainage of excess water, and extension of plant root networks.

Similarly, well-developed soil structure plays a crucial role in the reduction of erosion risk. Plant growth tends to be most effective in soils where crumb and granular structures predominate; optimum growth conditions are often associated with crumb aggregates of 1-10 mm diameter.

Soil structure should not be confused with the *tilth* of the soil. Provided the appropriate soil constituents are present, structure will develop in response to natural processes, such as wetting and drying and freeze-thaw cycles. In contrast, tilth is created by tillage operations, such as harrowing and discing, which break down large aggregates to form a finer seedbed suitable for the germination and emergence of crops.

The production of a good tilth requires great skill and practice. If the soil is too wet, then compaction may lead to impeded drainage and poor development of plant root networks.

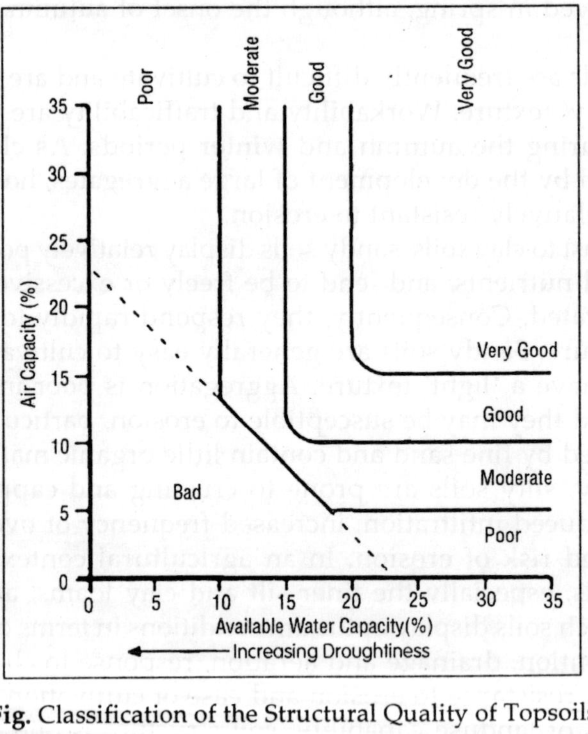

Fig. Classification of the Structural Quality of Topsoils

Furthermore, the resulting structural damage may increase the susceptibility of soils to erosion. These problems may be overcome to some extent through the adoption of minimum or zero-tillage options, although these are not suitable for all soil types and conditions.

The porosity and pore size distribution of soils are influenced directly by texture and degree of structural development. They are also closely associated with bulk density and have a major influence on water retention and aeration characteristics of soils. For example, if porosity is reduced due to compaction by heavy agricultural machinery, then increased waterlogging may lead to poor germination and emergence of crops. Restricted root network development may also lead to poor crop performance in such conditions. In fine-textured soils with naturally poor drainage, however, mechanised tillage operations may lead to improved drainage due to increases in macroporosity.

The beneficial effects of ploughing on soil structure and macroporosity in a farm-scale study of poorly drained clay soils in southern England. Here, macroporosity was found to be consistently higher in ploughed soils, even though they were traversed by agricultural machinery, than in direct-drilled soils where the use of such machinery was limited.

The content of plant-available water is a vitally important soil characteristic in relation to cropping potential. The amount of plant-available water depends largely on texture, porosity and pore size distribution, degree of structural development, and organic content, and is at its maximum in soils that are near field capacity.

In coarse-textured, sandy soils, amounts of plant-available water are small as water drains rapidly, under the influence of gravity, from relatively large pore spaces. Such soils tend to be drought susceptible (droughty) and irrigation water is often needed to supplement soil water reserves. In fine-textured, clayey soils, amounts of plant-available water are also small.

Here, much of the water is held in very small pore spaces at tensions greater than permanent wilting point (> 15 bars) and is thus unavailable to plants. Such soils tend to suffer from restricted drainage and are susceptible to waterlogging. In these circumstances, removal of excess water is encouraged through drainage practices. Amounts of plant-available water are greatest in soils with a silt loam texture.

When the water content of a soil is less than field capacity, it is said to be in a state of soil moisture deficit (SMD). The potential SMD is calculated on the basis of a growing season and represents the accumulated sum of potential evapotranspiration minus precipitation.

A cereal crop which yields 10 t ha^{-1} of dry matter, for example, could transpire 2,000-5,000 t ha^{-1} of water during a growing season, which is equivalent to 200-500 mm of rainfall. This value makes no allowance for drainage losses, surface runoff or evaporation directly from the soil surface.

Thus, in drier parts of eastern and southern Britain, for example, where mean annual rainfall is near 500 mm, crop yields may be adversely affected and irrigation is required to supplement soil water reserves on arable land. When the water content of a soil is above field capacity, cultivation using heavy machinery, and excessive animal stocking, may cause compaction, particularly in clay soils. In most instances, the field capacity period sets broad limits to good ground conditions for use of agricultural machinery and for animal stocking.

An understanding of the hydrological characteristics of soils is important not only on agricultural land, but also on land used for urban and industrial activities. For example, drainage and aeration problems are commonly experienced on land which has been restored following opencast mining programmes.

Usually, these problems are overcome through the installation of appropriate drainage systems. Similarly, the behaviour and migration of solutions are of critical importance in relation to contaminated land, and land which is used for waste disposal purposes.

The mechanical characteristics of soils have a major influence on their trafficability, and determine the nature and timing of mechanical operations. If agricultural machinery is used on soils which are wetter than their plastic limit, severe structural damage, compaction and the development of cultivation pans may occur. These problems will in turn result in impeded drainage and increased susceptibility to waterlogging, and under these conditions crop performance may be severely restricted.

Problems may also arise in soils with unfavourable shrink-swell characteristics. Expansion of swelling clays on wetting and shrinkage on drying result in churning and cracking of the soil. These changes cause disruption of soil structure and tilth, and are likely to have an adverse effect on crop yields. Vertisols in particular undergo swelling and shrinkage, in response to seasonal wetting and drying cycles, and are thus very difficult to cultivate.

In pastoral systems, the mechanical characteristics of soils have an important bearing on the timing and intensity of animal stocking. Once again, under wet conditions, particularly when the soil is above its plastic limit, animal trampling and associated compaction results in poaching and puddling of the soil.

Similar mechanical problems also occur in forested systems where heavy machinery is used in land preparation prior to planting, and during felling operations. In many areas of the world, forestry is practised on land which is unsuitable for arable cultivation. Soils are often developed on steep slopes, or are highly organic or have high water contents, and therefore tend to be mechanically weak.

Under these conditions, severe disturbance may result in

compaction, increased rates of organic matter breakdown and enhanced soil erosion.

On land used for urban and industrial activities, the mechanical characteristics of soils, particularly load-bearing capacity and consistency limits, have a major influence on the stability of buildings and their foundations. Similarly, if the substrate contains large amounts of swelling clays, stability may be adversely affected. This has been illustrated in many parts where, in recent years, drought has resulted in severe shrinkage of clay soils and parent materials, and consequent subsidence of buildings.

Soil temperature has a major influence on plant growth at all stages. At very high temperatures, however, the main limitation to growth is intense evaporation, and the resulting moisture stress, rather than temperature alone. Under these conditions, plants may suffer from physiological drought. Here, plants are able to cope with daytime moisture deficits, induced by intense evapotranspiration, by making best use of night-time moisture excesses.

Soil Biochemical Factors

Soil organic matter plays a fundamental role in agricultural land use, largely through its influence on water content, nutrient status and structural stability. The humus fraction in particular has a very high water-holding capacity and can retain up to four times its own dry weight of water; about 50 per cent of this is likely to be plant-available.

In fact, the presence of 5 per cent humus will increase the plant-available water content of a sandy loam by more than 50 per cent, and that of a clay loam by about 30 per cent, above levels in comparable organic-free soils. In addition to its water retention properties, organic matter is often added to soils as a mulch to reduce water losses by evaporation. Organic matter is also an important source of essential plant nutrients, particularly nitrogen.

The humus fraction in particular has a very high CEC and is therefore able to retain nutrients such as base cations, which are available for plant uptake. In a mechanical context, organic matter plays a crucial role in the development of soil structure. For this reason, soils with appreciable quantities of organic material are less susceptible to erosion than comparable soils with low organic contents.

The organic content of soils varies dramatically in response to differences in land use. Contents are often lowest in soils under intensive arable cultivation, as relatively little organic matter is returned to the soil after harvesting. Moreover, with the advent of inorganic fertilisers, the importance of organic material as a source of nutrients has declined. In southern and eastern areas of Britain, where arable cultivation is widespread, soil organic contents are often below 2 per cent by weight, although values vary somewhat depending on the type of crop rotation system adopted, and position within the rotation cycle. In soils under permanent pasture, levels of organic material are considerably greater than in arable soils, and are often in excess of 10-15 per cent.

Furthermore, soils under continuous arable cultivation often suffer a rapid decline in organic content over time, unlike those under pasture where organic contents remain relatively high. In forested soils, particularly those under needleleaf trees, organic contents can be very high.

In many parts of upland Britain, for example, needleleaf plantations are often established on soils which are peaty in character.

The size and diversity of soil organism populations are just as important as organic contents in terms of land use. Earthworms in particular are known to have a beneficial effect on nutrient levels, and on structural stability, in agricultural soils.

The number and diversity of soil organisms are often greatest in soils under permanent pasture where relatively high organic and moisture contents are maintained. In soils under arable cultivation, however, populations are often restricted as the food supply is relatively low. Similarly, in forested soils, particularly where needleleaf plantations are widespread, soil organism populations are restricted due to high levels of soil acidity and the prominence of unpalatable organic material.

Plant nutrient status is defined in terms of a number of soil properties, including exchangeable base content, nitrogen and phosphorus contents, CEC, per cent base saturation and acidity. In fertile, nutrient-rich soils, CEC is dominated by exchangeable bases and the percentage base saturation is high. In acid, nutrient-poor soils, however, CEC is dominated by H^+ and Al^{3+} and base saturation is low.

Here, it may be increased by liming, where adsorbed H^+ and Al^{3+} ions are replaced by Ca^{2+} ions. Similarly, levels of K^+ ions in the soil may be increased by the addition of NPK (nitrogen-phosphorus-potassium) fertilisers. It should be noted, however, that prolonged dominance of the soil exchange complex by one or two specific cations is detrimental to the availability of other cations to plants.

Such competition between different ions for plant uptake is known as 'antagonism'. High levels of Na^+, for example, are common in saline and sodic soils of semi-arid environments. Na^+ is toxic to many plants and tends to inhibit crop performance. Furthermore, high levels of Na^+

encourage aggregate dispersion and consequent breakdown of soil structure.

In addition to the plant macronutrients, amounts of soil micronutrients, or trace elements, such as molybdenum, boron and copper, may need to be controlled. Deficiencies or toxic excesses are often evident in crops and grazing animals, and can be regulated either through soil amendments or by direct treatment of crops and animals.

Many anions are just as important as cations, in plant nutrition. Such anions include nitrate, phosphate and various forms of sulphur. As AEC is restricted in most soils, particularly in temperate environments, anion retention is limited.

Consequently, losses through leaching tend to be rapid and levels which are sufficient for productive arable cultivation can only be maintained through fertiliser additions, or perhaps by mulching. The close relationship between nutrient status and organic content of soils, in a study of changes in the nitrogen content of arable soils in response to different manurial treatments; nitrogen contents were found to be considerably greater in soils which received larger inputs of farmyard manure (FYM).

Acidity plays a crucial role in the availability of plant nutrients. If pH is too high (> 8.0), or too low (<5.0), availability is likely to be restricted for certain nutrients. For most arable crops, growth is optimal at pH 6.0-7.0. Soil acidity is, however, rather unstable and susceptible to change. Excessive addition of nitrogen fertilisers, for example, may result in soil acidification.

Grassland vegetation on pastoral land, particularly that in humid, cool temperate environments, grows most effectively at soil pH values of 5.0-6.0, while needleleaf trees often grow most effectively where soil pH values are even lower (3.5-4.5); few other productive crops can tolerate this level of acidity.

Soil Type

Because soil type is defined by a combination of numerous soil characteristics, its influence on land use will be more general and less direct than that of an individual soil property, such as organic content, texture or acidity. Consequently, the relationship between soil type and land use is observed most clearly in geographical investigations of relatively large areas, particularly at the regional or national scale where a range of soil types and land uses may be identified.

At the more local scale, individual soil properties are likely to show much greater spatial variation than soil type, and therefore to have a much greater influence on land use. The broad geographical relationship between soil type and land use will be examined with reference to soil orders of the Soil Survey Staff classification system; approximate major soil grouping equivalents according to the FAO-Unesco classification.

Entisols

These are shallow, immature soils which are found in a variety of situations, including steeply sloping sites with colluvial parent materials, and floodplain sites on alluvial deposits. They may also exist in cold or exposed environments where soil formation is restricted. Entisols often have a low water-holding capacity and low nutrient status, and can rarely be used for productive arable cultivation.

They may be used for grazing although the quality of pasture is likely to be limited unless a pasture improvement programme is adopted. Entisols are rarely deep enough to sustain productive forestry, although shallow-rooted species may grow well in such soils.

Inceptisols

These are deeper and rather better developed than Entisols, although the morphological expression of soil development may not be particularly strong. They often have a relatively low nutrient status but with careful management they can be highly productive; this might include maintenance of organic matter levels and protection against erosion.

Inceptisols are widespread in their occurrence and are particularly common in floodplain areas of the River Ganges in India, and in many other parts of southern Asia where they are often used for high yielding rice production. They are also common in the Amazonian region of Brazil, and in parts of the USA they are used extensively for sugar-cane production. In the Sahelian zone of Africa, however, where Inceptisols are common, severe drought combined with widespread overgrazing has resulted in severe erosion problems.

Alfisols

These are particularly common in humid temperate environments and are widespread in western Europe, including eastern and southern areas of Britain, and in eastern parts of North America. Alfisols, particularly those with a fine loamy texture, are often highly fertile and contain significant quantities of mull humus.

They also have a high water-holding capacity and stable soil structure, and therefore tend to be well-suited to productive arable cultivation, scoring highly in terms of landuse capability. In Britain, they are often used for intensive cropping of both winter and spring cereals and exhibit few limitations to agriculture. In low-lying areas, however, particularly where profiles have a pronounced clay enriched B (argillic) horizon, waterlogging may present problems during the autumn and winter months.

In such circumstances, trafficking is restricted to prevent compaction and the formation of cultivation pans, and installation of artificial drainage is often necessary.

Mollisols

Mollisols are particularly common in relatively dry temperate environments, under prairie or steppe grasslands of continental interiors, such as central North America and Eurasia, and Argentina. These soils are rich in mull humus and have a well-developed crumb structure in their upper horizons. They also contain high levels of calcium and large earthworm populations, and are potentially very fertile.

They are, however, regularly exposed to drought conditions and, because they are often developed in loess parent materials, they tend to be susceptible to wind erosion, particularly if the surface organic-rich layer is degraded or removed. If carefully managed, using appropriate irrigation and crop rotation practices, Mollisols are highly productive and are widely used for arable cultivation in areas where population densities are relatively low.

In such areas, cereals, especially wheat, are widely grown; collectively, these areas constitute the world's 'bread basket'. In Argentina, however, extensive rearing of beef cattle is also practised in areas dominated by Mollisols.

Histosols

These soils may develop in a variety of situations from upland bog and moorland sites, to lowland fen and carr environments. Some Histosols are highly acidic and nutrient-poor, while others are nutrient-rich, and all tend to be waterlogged for long periods of time. Acid, nutrient-poor Histosols are of limited agricultural use, while nutrient-rich peats can be of great value for arable farming, once they have been drained.

However, drainage leads to rapid decomposition (oxidation) of organic matter which has a detrimental effect on structural stability, water-holding capacity and nutrient status. Furthermore, conservationists are particularly concerned about the destruction of wetland habitats, caused by extensive drainage of these soils.

Spodosols

These often have a peaty surface horizon consisting predominantly of mor humus and poorly decomposed plant remains. In addition, they are often coarse-textured, highly leached and acidic in character.

For the most part, Spodosols have a low water-holding capacity, which renders them drought susceptible, and also have a low nutrient status. Consequently, they are rarely used for productive agricultural purposes except in the case of forestry.

Spodosols are widespread in cool temperate, humid regions, such as the boreal forest zones of North America and Eurasia, particularly where soil parent materials are low in base-rich, weatherable minerals. In many parts of eastern USA, between the Great Lakes and the Atlantic seaboard, much of the virgin land colonised during the late eighteenth and early nineteenth centuries was underlain by Spodosols. This was of low fertility, and crop yields were so low that much of it had been abandoned by the mid-nineteenth century.

Ultisols

Ultisols are at the 'ultimate' or end stage of weathering and contain few weatherable minerals. These highly weathered and leached soils are acidic and of low nutrient status, and are dominated by kaolinitic clays. They are mature soils which have developed over a very long time period, in humid warm temperate to tropical environments. Ultisols occur most commonly in southeastern parts of the USA, eastern and southern parts of Brazil, and in parts of southeast Asia.

Although the inherent fertility of these soils is rather low, they can be utilised for productive arable cultivation if carefully managed. In southeast USA, Ultisols used to be subjected to a type of shifting cultivation regime, based largely on tobacco and cotton, the soils being cropped for a few years and then abandoned as fertility declined. However, this practice became unnecessary with the development of effective crop rotation systems which incorporated the use of lime, manures and legumes.

Oxisols

Oxisols are very highly weathered and leached, and have developed over long time periods. In addition, they are often acidic and of low nutrient status. They are characterised by high contents of oxides and hydroxides of iron and aluminium, and occur most commonly in humid tropical environments, particularly in Africa and South America.

In rainforest areas of central Africa and Amazonia, where rainfall is evenly distributed throughout the year, Oxisols are used largely for shifting cultivation. Here, small-scale plots of land are cleared for cultivation and then abandoned after a few years due to declining fertility and weed invasion.

The land then reverts to forest-fallow and nutrient levels are gradually restored. In a regional context, however, shifting cultivation can only sustain a relatively low population density. In savanna and savanna-woodland areas, where rainfall is more seasonal in character, Oxisols are not well suited to shifting cultivation, and pastoral activities are more common. Plots of land used for arable cultivation can then benefit from applications of manure derived from grazing animals. In spite of their low fertility, Oxisols can be highly productive with intensive management.

Vertisols

Vertisols are characterised by the presence of swelling clays and are subject to severe cracking on drying, and to swelling on wetting. Such physical disturbance often leads to tillage and trafficability problems, and to restricted water availability. In spite of these problems, however, these soils often have a high nutrient status and are highly fertile.

Vertisols occur most frequently in warm temperate to tropical environments, and are particularly common in Australia, India and Sudan. They also occur in southern coastal regions of the USA where they are developed in marine clays.

Cultivation of these soils is difficult due to their adverse physical characteristics. They tend to be very hard when dry and highly plastic when wet. Consequently, the timing of various cultivation operations is critical. In India, for example, planting of summer crops, which should ideally take place just before the monsoon rains, was virtually impossible in soil which was severely cracked and hardened following the long dry season.

Similarly, once the monsoon rains commenced, planting was also difficult as the soils soon developed plastic characteristics. Thus, planting had to take place just after the monsoon rains had finished. This practice was rather inefficient as the crops were unable to take full advantage of available water in the soil and depended primarily on the presence of stored water.

This problem of timing has now been overcome using a different approach to tillage, developed from research at the International Crop Research Institute of the Semi-Arid Tropics (ICRISAT). Soils were tilled immediately after the harvest of the winter crops, before the soil became too dry. Farmers could then establish the summer crop in loose dry soil just before the monsoon rains commenced.

Aridisols

These develop in arid and semi-arid environments, where soil water is severely restricted and productive agriculture is virtually impossible without irrigation. Poor management of Aridisols often results in soil degradation by wind erosion, salinisation and sodification. High levels of sodium in such soils are of particular concern. Sodium is toxic to plants, although the degree of tolerance varies from species to species. High levels of sodium can also lead to aggregate dispersion and consequent breakdown of soil structure.

Effective management of soil salinity and sodicity depends on sensible irrigation and drainage practices, together with evaporation control, selective chemical treatments, and the use of salt tolerant crops. With careful management, Aridisols are able to support a range of arable and pastoral activities. In the irrigated 'western range region' of the USA, for example, grazing of cattle and sheep is widely practised, together with cultivation of arable crops.

Andisols

These develop in parent materials of volcanic origin, notably ash and related deposits. They tend to have a relatively fine texture, with large amounts of fresh, weatherable minerals and low bulk density. They also contain significant quantities of organic matter and have a high ion exchange capacity. Consequently, they are often very fertile and tend to be intensively cultivated, particularly in tropical environments where they are renowned for their flexibility.

In some instances, however, aluminium saturation and low phosphate availability may present problems. Moreover, Andisols are commonly developed on rather steep slopes in mountainous regions, such as parts of southeast Asia, central America, east Africa and North Island New Zealand, and are fairly localised in their occurrence. Consequently, these soils may be particularly susceptible to erosion if poorly managed.

Additional Factors

A number of additional factors have an influence on landuse potential. These include climate, relief, parent material, biota and human activity. As these factors are closely interrelated, however, and because they act, in part, via the soil, evaluation of the effect of individual factors is rather difficult. Perhaps the most important factor is climate, which has a direct bearing on landuse capability. Climatic parameters include mean annual precipitation and potential evapotranspiration, which determine the level of plant-available water and extent of the soil moisture deficit (SMD). In Britain, for example, the 800 mm isohyet forms a distinctive boundary between predominantly arable cultivation in the drier east and south, and greater pastoral activity in the wetter west and north.

In terms of agricultural land use, the seasonality and intensity of precipitation are just as important as the mean annual total. Many areas of the world have a pronounced seasonal distribution of rainfall, notably the monsoon areas of southern Asia, savanna regions of Africa, Mediterranean countries and western USA.

These have a prolonged dry season when SMD is a problem and irrigation is often necessary. Rainfall is concentrated into a few months of the year and is often very intense. If agricultural land is not carefully managed, or the planting of crops is badly timed, this type of rainfall can be particularly destructive, resulting in severe soil erosion and consequent loss of productivity.

In many parts of India, for example, planting of the summer crops is timed to coincide with the pre-monsoon rains so that a protective vegetation cover is established before the main monsoon begins. In many cool temperate areas, soils are considerably wetter during the winter months than at any other time of year. Under these conditions, workability and trafficability are severely restricted. If soils remain wet during the spring and early summer periods, many crops become increasingly susceptible to pests and diseases.

In addition to precipitation, temperature plays a crucial role in agricultural landuse capability. Temperature has a direct influence on growing season length, and on rates of crop growth and maturation. Although mean annual temperature is important in this respect, seasonal variations in temperature are often more significant, particularly in temperate regions; in tropical environments temperatures are always high, except in mountainous areas, and plant growth continues throughout the year.

The accumulated temperature above 0°C for the first six months of the year (January-June) is frequently used as a measure of heat energy available for plant growth. Average (median) values of this index appear to correlate well with the length of growing season. For some crops, however, the more usual index of accumulated temperature above 5.6°C is preferred. Another important climatic parameter, closely allied to temperature, is incidence of frost.

Although frost plays an important role in the development of soil structure, and in the control of pests and diseases, it can present problems if it occurs in the early stages of crop growth. Fruit trees in particular are very sensitive to late frost during the spring when they are in blossom. Similarly, grape crops are sensitive to early frost during late summer and early autumn. This may inhibit ripening of the fruit. which tends to swell and split.

In addition to climate, relief has a major impact on landuse activities. Important parameters include altitude and aspect, both of which have an influence on climate, and slope angle. Generally, limitations on land use increase with increasing altitude, particularly in temperate environments, where pastoral activities replace arable as conditions become cooler and wetter.

The effect of aspect is felt most strongly in hilly and mountainous terrain of high latitude areas. Here, the angle of incidence of the sun is relatively low for much of the year and large areas may be shaded. Thus, in the northern hemisphere, soils of north-facing slopes tend to be colder and wetter than those of south-facing slopes, and the duration of snow cover in winter is very much greater.

In areas where the climate is already cool and wet, unfavourable aspect may severely restrict landuse options. At lower latitudes, however, where the climate is warmer, soils on a well-shaded slope often have higher moisture and organic contents than comparable soils on slopes which are exposed to direct sunlight. Here, evaporation is intense and organic matter is rapidly broken down, thus restricting landuse options at such sites.

One of the most important limiting factors in terms of land use is slope angle. This has a major influence on soil depth and stability, and on workability of the land. On steep slopes, soils are often shallow and unstable, and are particularly susceptible to erosion. This problem is compounded by the use of agricultural machinery. On low-angled slopes, which are less than about 10°, land is usually cultivated parallel to the contours, thus reducing the risk of soil erosion. On steeper slopes, however, agricultural machinery is usually operated perpendicular to the contours in order to minimise the risk of overturning.

It is well known that erosion risk increases dramatically on land which is cultivated in this way, as overland flow is concentrated by compacted wheel tracks and plough furrows. Slope angle also has an important bearing on soil conservation practices. As slope angle increases above about 10°, contour cultivation practices such as strip-cropping and contour bunding are usually replaced by terracing. If slopes steeper than about 30° are terraced, however, the risk of land slippage increases dramatically, thus posing a threat to property, communication systems and life itself. Parent material has an effect on land use, although this operates via the soil and is thus less direct in its influence than climate and relief. Soil characteristics which are most strongly associated with parent material include stone content, texture, mineral nutrient levels and acidity.

Vegetation and soil organisms existing prior to cultivation may give an indication of landuse potential. Good quality grassland and deciduous woodland, for example, indicate soils of relatively high nutrient status. Such soils usually have high earthworm populations and significant quantities of mull humus, and if they are well drained, they could be used for productive arable cultivation.

However, the rapid expansion and intensification of arable cultivation in recent decades has led to overproduction, particularly in the developed world, and to soil degradation, often at the expense of areas with natural and semi-natural vegetation covers. Thus, conservation of natural and semi-natural vegetation, and the re-establishment of such vegetation in formerly cultivated areas, are often encouraged in preference to clearance and cultivation.

Today, many of the physical limitations on land use can be overcome, to some extent, by human intervention, particularly in agricultural systems. Since settled agriculture began, several thousand years ago in some parts of the world, there have been tremendous developments in crop rotation practices, seed and crop varieties, levels of mechanisation, tillage practices and agricultural chemicals.

These developments have been coupled with advances in our understanding of soil and water conservation, drainage and irrigation, and pollution control, and all of these factors have had a dramatic impact on the performance and efficiency of agricultural systems, particularly in the developed world. Many would argue, however, that intensification of agriculture and increased productivity are not sustainable in the long term, and are likely to have an adverse effect on the environment.

In addition to the above, a number of other factors have an important bearing on the operation of landuse systems. These include economic considerations, the nature of land tenure, farm policies, political and infrastructural support systems (e.g. transport, communications, education and incentive schemes), and social and cultural structures.

Their impact can perhaps best be appreciated through consideration of the dramatically contrasting situations in the developed and developing world. In the developed world, populations are relatively stable in terms of their size and spatial distribution.

There is also more productive land per head of population and a wider choice of landuse options than in most developing countries. These factors alone often ensure the most effective and appropriate use of land. Furthermore, the relatively high economic status of land users in developed countries facilitates the establishment of high input agricultural systems which utilise high yielding crop varieties, are highly mechanised, and which rely on a range of agricultural chemicals.

Land users are also able to invest in long-term conservation strategies, a practice which is often supported at national level by government policies, such as the 'Set Aside' scheme in Britain. Here, land users are given a financial incentive to plant trees and hedgerows and to preserve various endangered habitats by setting aside land formerly used for productive arable cultivation. Not only does this contribute to environmental conservation, but it also helps to reduce excessive agricultural production, and the cost of maintaining excessive food reserves.

In the developing world, population is growing rapidly and is often unstable in terms of its spatial distribution. Rural-urban migration is widespread and the growing number of poor, peasant farmers are forced onto increasingly marginal land which is unable to sustain intensive use. In other areas, exhausted land is abandoned and becomes severely degraded.

Extensive exploitation of woodfuel resources in many developing countries has also lead to widespread land degradation. In extreme circumstances, famine and warfare have resulted in large-scale migration with consequent increases in refugee populations, particularly in parts of Africa. The majority of land users suffer extreme poverty and are forced to eke out a living from tiny small-holdings on land which they do not own.

Peasant farmers are driven by short-term needs rather than by long-term goals. Rarely can they afford to invest in high input agricultural practices which require purchase of expensive seed, fertilisers and pesticides, and the use of agricultural machinery. In fact, such practices and associated technology are inappropriate in many developing countries, where agriculture is often particularly labour intensive.

Instead, farmers have to be resourceful and must rely on cheaper alternatives which are closer to hand; these include animal manure, fire ash, crop residues and household waste. In addition, poverty and the small-scale of many landuse operations do not allow establishment of large-scale, elaborate conservation schemes like those often found in the developed world.

AGRICULTURAL LAND USE

Arable Cultivation

Methods of arable cultivation have evolved in different ways in response to the vast array of contrasting global environments. In many humid tropical regions, where soils are of low fertility, shifting cultivation, sometimes known as *slash and burn,* is commonly practised, although this often sustains only a small fraction of the population in such regions. Essentially, shifting cultivation involves a three-phase rotation.

This commences with the clearance of existing vegetation, often by burning which helps to improve soil nutrient status, particularly in tropical ecosystems where much of the nutrient stock is stored in the biomass rather than in the soil. Following clearance, the land is cultivated, but for only a few cropping cycles (cultivation phase), as the decline in soil fertility is rapid and weed invasion is a major problem.

The period of cultivation is influenced to some extent by rainfall, with wetter areas being cropped for shorter periods than drier areas. Generally, however, a cropping period of about one to five years is commonly adopted. After cultivation, the land is abandoned and former vegetation is allowed to re-colonise (bush-fallow phase). This phase is essential for the restoration of soil nutrient levels, and its length again depends to a large extent on rainfall. Usually, the recommended period is five to twenty years, although it may be slightly longer in wetter areas and a little shorter in drier areas.

An example of shifting cultivation is the Chitemene system of northern Zambia. Here, felling of trees is limited and crops are grown in ash gardens created by burning lopped branches. Similarly, the Jhum system practised in Bengal, India comprises five-, ten- and twenty-year cycles with a long fallow and a variety of mixed cropping. A related practice is the Khoriya system of Nepal.

In temperate areas, land has often been more closely settled, and competition for land has been greater, than in tropical regions. Furthermore, land in temperate areas cannot be cropped for significant periods of the year, due to the restricted growing season, and consequently more intensive cropping systems have evolved.

One of the oldest recorded systems in Britain, practised as early as the Celtic period, was a three-year rotation of autumn cereal, spring cereal and fallow. This was largely replaced during the eighteenth century by other rotations such as the Norfolk four-course system. Here, grass-clover pasture, root crops and cereals were alternated, and arable practices were closely integrated with livestock rearing. The nitrogen for crop growth could therefore be fixed in the soil by clover, which is a legume, and released from grass residues and manure.

During the nineteenth century, other cropping practices evolved, largely in response to depressed cereal prices and the increasingly accepted belief in the beneficial effects of grass as a soil 'conditioner'. At this time arable-pasture (or arable-ley) rotations, with a relatively long period of ley (up to three years), became increasingly common. It was during this period that mixed farming became widespread, with livestock being reared on farm-produced grass, cereals and root crops.

This type of farming was based on a nutrient supply from farm-derived organic materials such as farmyard manure (FYM), which consists of animal dung and urine mixed with straw, slurry (dung suspended in urine and washing water) and compost. Sewage sludge from local sewage treatment works has also been used on agricultural land. Unfortunately, however, this material often contains high levels of heavy metals, such as lead, zinc and cadmium, derived from industrial effluent.

Green manure crops, sometimes known as *catch crops*, have been an additional source of nutrients. Here, a quick-growing leafy crop is ploughed into the soil, releasing nitrate for the benefit of the following cash crop. This approach reduces the risk of nitrate leaching, but its effects are short-lived and it contributes little to soil organic matter levels.

In recent decades, due largely to increasing population pressure, demand for food and competition between different land uses, there has been a move away from the more traditional agricultural practices outlined, in both tropical and temperate regions. During the Second World War in particular, the demand for food increased and agricultural activity intensified.

Arable and livestock rearing became increasingly separated so that FYM was no longer available where it was needed. At this time, greater emphasis was placed on cereal monoculture at the expense of arable-ley rotations. This change became feasible as the traditional sources of nitrogen were replaced by inorganic nitrogen fertilisers.

However, excessive use of such fertilisers led to increased soil acidification, and to nutrient enrichment of runoff from agricultural land with consequences for surface water quality and aquatic biota.

In addition to the chemical changes, the intensity and scale of mechanisation in arable cultivation has increased dramatically. This has contributed to increased soil compaction which in turn has led to impeded drainage and restricted development of plant root networks.

In an attempt to overcome these problems, minimum and zero-tillage options have been developed, where the use of heavy agricultural machinery is minimised.

An example of such a practice is direct drilling where crops are sown directly into the existing vegetation cover. This is now practised in many parts of the USA, particularly in the central corn/maize belt, and in many parts of tropical Africa, South America and Australia.

The advantages of minimum and zero-tillage include improved topsoil organic content, aggregate stability, moisture retention, erosion control and workability. In addition, energy requirements are reduced and excessive heating of the soil in hot environments is prevented. In a comparison of zero-, minimum and conventional tillage practices in northern Italy, Expert reported an improvement in a number of soil characteristics, particularly those relating to aggregate stability, with reduced tillage.

However, the disadvantages of minimum and zero-tillage include increased bulk density, reduced total porosity and slow warming of the soil during the spring. Moreover, phosphorus and potassium become concentrated in the surface layers (0-5 cm) of the soil which may restrict their availability to plants if the soil dries out. This is in contrast to tilled soils where the nutrients are more evenly mixed throughout the topsoil (0-20 cm).

In addition, accumulation of volatile fatty acids around fermenting crop residues may inhibit seed germination, and the increase in perennial weeds is difficult to control with herbicides alone.

In many tropical and subtropical areas, particularly in the developing world, low external input agriculture is becoming increasingly common as a component of sustainable development. Here, farmers use crop varieties which are adapted to the main soil constraints, while minimising the use of purchased inputs, and maximising nutrient recycling.

This is very different from the high input systems from which much of the world's food supply is derived, but which involve intensive use of agricultural chemicals to overcome soil constraints. This system was developed as part of the transition process between shifting and continuous cultivation.

The principal features include slash and burn clearance, rotation of acid-tolerant upland rice and cowpea cultivars, maximum residue retention, zero-tillage, and absence of lime and fertilisers. As yields declined, due to nutrient deficiencies and increased weed pressure, a kudzu fallow was grown for one year.

Subsequent options included pasture, agroforestry and fertiliser based continuous cultivation. It was suggested that this system helped to preserve some agro-ecosystem diversity, while contributing towards sustainable production and income for farmers.

All of the developments are clearly reflected in increased crop yields, particularly in the developed world. In western Europe, for example, the amount of grain produced per head of population increased from about 250 kg in 1950 to nearly 500 kg in 1990, an increase of 2.5 per cent a^{-1}.

To a large extent, these increases have been achieved through the intensification of agriculture, rather than by increasing the area of land under cultivation. In the developing world, however, production has increased only slightly, often by increasing the area of marginal land under cultivation, and in some areas production has declined.

In Africa, for example, the amount of grain produced per head of population decreased from approximately 150 kg in 1950 to about 120 kg in 1990, a decrease of 1.6 per cent a^{-1}. This decrease, coupled with rapid population growth and the extension of cultivation onto marginal land in many areas, is clearly not sustainable in the long term.

Pastoral Activities

Grazing as a landuse practice involves the rearing of animals mainly for transport, meat, milk or wool. In many areas, natural and semi-natural grassland and scrub are used for grazing. These include parts of the prairies and steppes of the temperate continental interiors of North America and Eurasia, the pampas of South America, and the savanna and rangeland areas of Africa and Australia.

Even some of the most fragile ecosystems are used to support grazing livestock, including semi-arid regions of Africa and Asia, and tundra areas of northern Scandinavia.

In other parts of the world, natural and semi-natural vegetation has been cleared or at least substantially modified to allow good quality grassland to become established. The quantity and quality of herbage can be improved either by replacement of existing vegetation, in conjunction with fertiliser and lime treatments, or by improvement of existing grasses.

Although the former approach is often considered to be the most effective, it takes longer to achieve and is more costly than the latter. Once grazing has commenced, re-establishment of pre-existing vegetation is prevented by the consumption of new shoots.

Until the 1960s, almost all hill land improvement in Britain involved ploughing and reseeding, which proved to be appropriate for the brown earth and podzolic soils most commonly associated with grazing prior to this time. More recently, however, with the extension of pasture improvement onto peaty gleyed podzols and humic gleys, the more gradual improvement of existing grasses appeared to be a more appropriate option.

Consequently, in order to achieve the most effective solution, in economic and biological terms, an understanding of soils, vegetation and grazing strategy is essential. One of the most widely used pasture improvement strategies in the British uplands still involves the replacement option. Soil pH is raised to about 5.5 or more by liming.

Soil nutrient levels are also increased by fertiliser treatments and a seed mixture of grasses and white clover. Clover, being a legume, helps to fix nitrogen in the soil, a process which is enhanced by inoculation of clover seed with nitrogen-fixing bacteria such as *Rhizobium* spp.

Sowing is usually carried out in April or May, with limited cultivation and light grazing initially. Grazing can also be controlled effectively using appropriate fencing strategies. Pasture quality must be maintained through the application of dressings of lime and fertiliser, particularly at wetter sites. In addition, livestock should be monitored for problems relating to trace element deficiency or toxicity.

A number of environmental problems have been associated with pastoral activities, including erosion, compaction, salinisation and acidification of soils, and leaching of organic wastes into potable water supplies.

The decline in soil organic content and nutrient levels in the grazed plots was explained in terms of a number of interrelated causal factors. Removal of the protective vegetation cover by grazing livestock and by burning, together with soil compaction caused by animal trampling, resulted in increased soil exposure and enhanced surface runoff.

These increases in turn resulted in accelerated soil erosion, organic matter decomposition and leaching losses. It appeared that the return of organic matter and nutrients to the soil by grazing livestock was insufficient to counteract the deterioration in soil herbage quality. Suggested responses to these problems included rotational grazing, incorporating the planting of a supplementary fodder of legumes in grazing areas. This practice may reduce the frequency of overgrazing, particularly during the dry season.

Hay storage for use during the latter part of the dry season was also encouraged. In addition, it was suggested that the timing and amount of burning should be reviewed. Burning helps to reduce pest levels and to restrict bush invasion, but it also destroys soil organic matter. It was suggested that burning should be carried out early in the dry season, before the ground becomes tinder dry, rather than late in the season when it can be particularly intense and destructive.

Another problem with grazing in dryland environments is that it has traditionally been transhumant, with pastoralists moving to floodplain areas in dry seasons and away from them in wet seasons. Commercial irrigated farming in floodplain areas has prevented this, leading to the over-use of other, more marginal areas for grazing.

Forestry

Many areas under natural and semi-natural forest and woodland have been, and are currently being, exploited for their timber, often using large-scale, mechanised logging operations. Such areas include the tropical forests of central Africa, Amazonia and southeast Asia, and the boreal forests of North America and Eurasia. It should be noted, however, that reliable estimates of deforestation rates are difficult to establish.

In recent years, satellite imagery has been used to assess the extent of deforestation, particularly in tropical forest areas. Ideally, this approach should be used in conjunction with ground verification otherwise estimates may be misleading and possibly erroneous. For example, areas of deforestation have sometimes been determined from aerial estimates of recent fire damage.

However, this can lead to overestimates of deforestation rates, as smoke plumes from recent fires are likely to cover a far larger area than that which has been deforested. Conversely, underestimates may result if forest areas are thinned rather than clear-felled; in this situation low or intermediate canopy trees may be removed, while the upper canopy remains intact.

The reasons for deforestation are largely economic. Softwoods from areas under boreal forest are used in the paper and construction industries, while tropical hardwoods are used in the manufacture of furniture and for construction purposes. In many developing countries, forest and woodland areas are also heavily exploited for woodfuel, which is the main source of energy, and for agricultural land.

Clearly, the exploitative decline in forest and woodland resources, is not sustainable in the long term and a number of environmental problems, including soil erosion, salinisation and biodiversity losses, have developed as a result. In addition, improved road access into deforested areas is likely to encourage population influx and increased shifting cultivation. A variety of approaches to these problems have been suggested.

These focus on the sustainable use of forests, which involves selective logging of several areas each year, followed by forest closure and regrowth. This approach allows for sustainable timber production, conservation of soil and water resources, and maintenance of ecosystem diversity. It also ensures the survival of indigenous populations within forest areas.

An example of sustainable forest management is agroforestry, where trees and arable crops are combined. This has been particularly successful in tropical and subtropical regions where the degree of shading is less than in temperate areas. The trees are usually planted in rows, between which a variety of crops are established.

Although the trees may compete with crops for nutrients, there are a number of benefits associated with the incorporation of trees into cropping systems. As well as providing shelter, trees help to control runoff and erosion. They also help to improve soil nutrient status through the addition of organic material and maintenance of soil structure. Moreover, the trees are an important source of timber, woodfuel, animal fodder and fruit crops.

The most successful agroforestry projects use indigenous tree species which are best adapted to local soil and climatic conditions. In India, for example, species of *Leucaena* are considered to be particularly beneficial as they are able to fix nitrogen in the soil. In contrast, eucalyptus trees are less favoured because they have a high water demand and their leaf litter is toxic.

Studies by the Soil Conservation Institute of the Indian Council of Agricultural Research have shown that eucalyptus trees planted close to crops may reduce crop yields by 10-15 per cent. In spite of these problems, eucalyptus trees grow rapidly, providing a useful source of woodfuel, and consequently they are often incorporated into agroforestry systems, but are usually planted in stands well away from crops.

Another approach to sustainable forestry and woodland management is the promotion of small-scale social forestry programmes, where local communities are involved in the establishment of tree nurseries. Farmers are then encouraged to transplant saplings from the nurseries onto their own land. Multipurpose tree species, which provide animal fodder, fruit and woodfuel, and which grow rapidly, are particularly favoured.

Since the late 1970s, social forestry schemes, established at the local *panchayat* (village lands) scale, have been successful in parts of Nepal, where they have been supported by the FAO, UN Development Programme, World Bank and Community Forestry Development Programme (CFDP). Sensible forestry and woodland management has also been facilitated by switching energy sources from woodfuel to cheap and sustainable alternatives such as bio-gas, which is derived from the decomposition of animal and human excrement and related waste products.

In many temperate regions, forestry programmes often utilise rapidly growing softwood species. Here, forestry practices are usually relegated to steeply sloping uplands which are of limited agricultural value. In the last few decades, many areas of the British uplands have been converted from moorland and rough pasture to needleleaf forestry.

The main species planted include Sitka spruce *(Picea sitchensis)*, Scots pine *(Pinus sylvestris)*, Lodge-pole pine *(Pinus contorta)*, Japanese larch *(Larix kaempferi)* and Norway spruce *(Picea abies)*. These species are well adapted to the cool climatic conditions, poor soils and steep slopes often found in upland areas, and are ready for harvesting within a few decades.

Although expansion of the forestry industry may be appropriate in economic terms, the environmental consequences may be serious, especially if expansion is poorly managed. In particular, coniferous afforestation has been closely associated with soil and surface water acidification.

As the soils in many planted areas are often wet and acidic, drainage and nutrient amendments are frequently required before planting can commence. Drainage ditches are usually constructed parallel to the direction of slope, and frequently feed into deep check drains which run across the slope. These serve to reduce rates of water flow and soil transport into adjacent streams.

Following these land pre-treatments, saplings are planted on the ridges between drainage ditches and fertilisers may be added to stimulate growth in the important early phase of tree development. About ten to twenty years later, thinning takes place. This practice encourages stronger growth, thus reducing the susceptibility of plantations to damage by windthrow which is a serious problem in exposed areas.

The trees are usually harvested after about forty to sixty years, towards the end of the optimum growth phase. After this, the economic rate of return diminishes as growth rate slows. The harvesting process is highly mechanised and may be carried out in one of two main ways.

First, only the trunk of the tree is removed, whilst the thin top and branches (brash) are left to protect the soil (lop and top felling). Second, all parts of the tree are removed (whole-tree harvesting) and the soil is left with minimum protection. In a study of whole-tree clear-cutting on soils it was observed increases in bulk density caused by mechanical compaction; for high-value timber, however, this can be alleviated by helicopter logging.

Decreases in pH, CEC and per cent base saturation were also observed, particularly in the upper soil horizons. In a similar study, compared the effects of conventional clear-felling (lop and top) and whole-tree harvesting of Sitka spruce on future productivity of forestry at a site in North Wales. Soil potassium reserves were used as an indicator of productivity, and although both methods of felling resulted in the return of some potassium to the soil, amounts were greater following conventional clear-felling.

NON-AGRICULTURAL LAND USE

Urban and Industrial Activities

Urban and industrial land use both influence, and are influenced by, the soil and its parent material. For example, construction and related engineering programmes are determined to a greater or lesser extent by the physical and mechanical characteristics of soils and their substrates.

Such characteristics include load-bearing capacity, shear strength, consistency, shrink-swell potential and drainage capacity. These parameters will influence the strength and stability of foundations, as well as the cost of construction operations.

Another aspect of urbanisation and industrialisation is waste disposal, much of which takes place at landfill sites. Seepage and diffusion of toxic liquids and gases may lead to soil contamination at such sites. The behaviour of toxic constituents in the soil will depend to a large extent on its physical characteristics.

Variations in texture, porosity and pore size distribution, in particular, will influence rates of liquid seepage and gaseous diffusion. These processes will also be influenced by the specific site characteristics and management techniques adopted at the landfill site. In addition to the burial of waste materials at designated landfill sites, surface disposal of low-toxicity, biodegradable waste is commonly practised.

An example of this is the application of sewage sludge, from sewage treatment works, onto agricultural land on the outskirts of urban areas. In spite of its organic- and nutrient-rich properties, sewage sludge is often found to contain high levels of heavy metals, such as lead, zinc and cadmium, derived from industrial effluent. This can result in the contamination of soils, and of the produce grown on them.

In response to this problem, the quality of sewage sludge must now be assessed before it can be applied to agricultural land.

In addition to land-based disposal, aerial dispersion of pollutants, and subsequent contamination of soils, is a major problem in many urban and industrial areas. Such pollution is not necessarily confined to areas immediately adjacent to centres of urban and industrial development, but may be transferred several hundreds of kilometres, often traversing national boundaries.

Sources of atmospheric pollution which may lead to soil contamination include acid emissions derived from the burning of fossil fuels, and radionuclide emissions derived from nuclear accidents. Other airborne contaminants often found in soils include organic compounds such as dioxins. A major problem with regard to the management of contaminated land is the absence of a consistent set of guidelines for threshold levels of the many and varied pollutants.

Another area in which soils are relevant to urban and industrial landscapes is that of land reclamation and restoration. This applies particularly to sites which have been used for opencast mineral extraction. Prior to excavation at such sites, topsoil and subsoil are usually removed and stored temporarily.

Storage often takes the form of banks or baffles which serve to hide the excavations, and to reduce levels of noise and dust. Once extraction is complete, the site is infilled with overburden, and subsoil and topsoil are then replaced. In some instances, however, there is insufficient bulk to fill the excavations, therefore landfill may be a viable option.

The main problems experienced with soil restoration at opencast sites include lack of aeration in stored soil mounds, and compaction by machinery on replacement of the soil. Erosion may also occur in the absence of a protective vegetation cover, particularly if the topsoil is poorly reinstated. Once the subsoil and topsoil have been replaced, applications of lime and fertilisers are often required, in combination with sensible tillage and land husbandry practices, and drainage systems need to be installed before conditions become suitable for effective plant growth. In a study of the cropping potential of restored soils at a site it was concluded that the installation of an effective drainage scheme was particularly important.

Here, yields of winter wheat and spring barley were found to be 30-40 per cent lower on undrained land than on drained land.

Recreation and Leisure

In the developed world in particular, the amount of time available for outdoor recreational and leisure activities has increased enormously in recent decades. To a greater or lesser extent, all of these activities put pressure on soils and ecosystems, and if they are poorly managed, land degradation may result.

This can occur, for example, in the form of footpath erosion, which is a problem on many long-distance footpaths. Also, in many upland and mountain areas, increased pressure on ski slopes has caused severe damage to rare and sensitive alpine plant communities. In Scotland, for example, damage to alpine vegetation communities has been particularly severe in parts of the Cairngorm area, and access has been restricted at a number of sites to encourage recolonisation.

Expansion of tourism in parts of the developing world has had an adverse impact on soils and ecosystems. In the Himalayan foothills of Nepal, for example, the growing number of trekkers has put pressure on already dwindling woodfuel resources through their demand for lighting, hot water, and other luxuries which they take for granted in their home countries. A similar argument may be applied to safari tourism in east Africa. As indicated earlier, exploitation of woodfuel has been associated with increased rates of soil degradation in many developing countries.

Another, more local aspect of recreation and leisure in which soils are important is that of sportsfield design and construction. In recent years, a considerable amount of research has emerged regarding the construction of artificial soil profiles on sports fields.

Important requirements for such profiles include a durable turf surface, a subsoil with high compressive and shear strength, and adequate drainage.

Chapter 13

Environmental Problems in Agriculture

All landuse activities, particularly those which are poorly managed, involve destruction or disturbance, to a greater or lesser extent, of natural and semi-natural ecosystems. Almost invariably, however, it is these ecosystems, in equilibrium with their environment, which offer most effective protection to the soil which supports them. A major consequence of ecosystem destruction and disturbance is that of soil degradation.

This has been defined as the decline in soil quality caused through its misuse by human activity. More specifically, it refers to the decline in soil productivity through adverse changes in nutrient status, organic matter, structural stability and concentrations of electrolytes and toxic chemicals. Soil degradation incorporates a number of environmental problems, some of which are interrelated, including erosion, compaction, water excess and deficit, acidification, salinisation and sodification, and toxic accumulation of agricultural chemicals and urban/industrial pollutants.

In many instances, these have led to a serious decline in soil quality and productivity, and it is only in recent decades that the finite nature of soil as a resource has become widely recognised.

Soil degradation is not a new phenomenon. Archaeological evidence suggests that it has been on-going since the beginning of settled agriculture several thousand years ago. The decline of many ancient civilisations, including the Mesopotamians of the Tigris and Euphrates valleys in Iraq, the Harappans of the Indus valley in Pakistan and the Mayans of Central America, was due in part to soil degradation.

More recently, an event of major significance was the dustbowl which occurred in the Great Plains of the American midwest during the 1930s. At this time, intensive agricultural practices, employed in the eastern states, were transferred to the drier midwest where the soils are lighter textured and more susceptible to erosion.

A number of years of drought, combined with crop failure and destruction of the protective organic-rich topsoil, resulted in severe wind erosion. According to the Global Assessment of Soil Degradation project, about 15 per cent of the global land area between 72°N and 57°S is degraded. Of this, an area slightly less than that of India (*c.* 300 million hectares) is strongly degraded, largely as a result of deforestation (113 million hectares), inappropriate management of cropped land (83 million hectares) and overgrazing (75 million hectares).

In recent decades the global rate of soil degradation has increased dramatically, and is likely to increase further as we approach the twenty-first century; in 1983 it was estimated at 5-7 million ha a^{-1} and is set to rise beyond 10 million ha a^{-1} by the year 2000.

The effects of soil degradation are not restricted to the soil alone, but have a number of off-site implications. Soil erosion, for example, is often associated with increased incidence of flooding, siltation of rivers, lakes and reservoirs, and deposition of material in low-lying areas.

These problems may be compounded in areas where infiltration capacity is reduced due to compaction, hardsetting or induration of soils. Salinisation and sodification of soils are often associated with poor quality irrigation water, while soil acidification is commonly linked with acidification and aluminium contamination of surface waters. Leaching of fertilisers and pesticides from agricultural soils may also lead to contamination of surface and shallow ground waters.

In addition, contamination of soils by urban and industrial pollutants, such as heavy metals and radionuclides, may lead to toxic accumulation in arable produce and in herbage for grazing animals, thus having important implications for human health.

The extent of soil degradation is influenced by a number of factors, many of which are interrelated, namely soil characteristics, relief, climate, land use, and socio-economic and political controls. In many studies of soil degradation and its wider environmental implications, the socio economic and political controls are often overlooked, or at least not examined in any detail, perhaps because of the difficulties associated with the collection of reliable and comparable data. Increasingly, however, these controls on landuse systems are being viewed as central to the issue of soil degradation, particularly in the developing world.

Management of soil degradation, whether at a global, regional or local scale, is clearly a complex issue and represents one of our most challenging environmental problems. Emphasis should be placed on sustainable rather than exploitative landuse practices; this theme was highlighted by the World Soil Charter which called for a commitment by governments, agencies and land users to 'manage the land for long term advantage rather than short term expediency'. The problem requires a holistic, multidisciplinary approach involving the collaborative and co-ordinated efforts of ecologists, agronomists, soil scientists, hydrologists, engineers, sociologists and economists.

Moreover, the involvement of government and non-government organisations, aid agencies and the farmers themselves is essential to the success of research and development in this area. Such involvement should facilitate the implementation of education, training and incentive programmes. Imposition from of high-technology, high-cost solutions by technical experts from developed countries is certainly not the answer in the developing world.

Inevitably such solutions are not economically viable and low-technology, low-cost options, such as low external input agriculture, agroforestry and social forestry, are often the only answer. Hence, the approach to soil conservation has shifted in recent years from a rather technocentric standpoint to a more ecocentric position.

Central to this approach are the concepts of land husbandry and sustainable development, which place emphasis on the land users themselves rather than on the technical experts and advisors.

A selection of the most pressing soil degradation problems, and in each case the causal factors, on- and off-site effects, and management strategies will be considered.

PHYSICAL PROBLEMS

Erosion

Soil erosion occurs when the rate of removal of soil by water and/or wind exceeds the rate of soil formation. Generally, rates of soil formation are very low, with profiles developing at a rate of about 1 cm every 100-400 years; assuming an average bulk density of 1.33 g cm^{-3}, this equates to about 0.3-1.3 t ha^{-1} a^{-1}.

It is important to differentiate between natural or background erosion, and erosion which has been accelerated largely as a result of human activity. Background erosion rates are often similar to rates of soil formation at <1.0 t ha^{-1} a^{-1}, although in mountainous areas they may be considerably higher. In contrast, rates of accelerated erosion commonly exceed 10 t ha^{-1} a^{-1} and sometimes exceed 100 t ha^{-1} a^{-1}.

Some of the highest soil erosion rates have been observed in the loess plateau area of China and in the Himalayan foothills of Nepal, where values in excess of 200 t ha^{-1} a^{-1} have been recorded. Similarly, in India, gully erosion results in a loss of about 8,000 ha of land per year. The extent of soil erosion is governed by a number of factors. Those of particular importance include erosivity of the eroding agent, erodibility of the soil, slope steepness and length, landuse practices and conservation strategies. These factors are summarised in the Universal Soil Loss Equation which has been used widely in the modelling and prediction of soil erosion:

Where E = mean annual soil loss, R = rainfall erosivity index, K = soil erodibility index, L = slope length, S = slope steepness, C = cropping factor which represents the ratio of soil loss under a given crop to that from bare soil, and P = conservation practice factor which represents the ratio of soil loss where contouring and strip-cropping are practised to that where they are not. Although widely used, this model has been the subject of extensive criticism.

For example, it assumes that a vegetation cover is always protective which is not necessarily the case; erosion on land with a good cover of crops planted in rows can be greater than on land which is sparsely vegetated. It is also water erosion based and cannot be used in areas affected extensively by wind erosion. More specifically, it focuses on rill and inter-rill erosion and is not easily applied to areas where gully and stream bank erosion are widespread. Its universal nature has also been questioned, particularly in terms of its application to tropical soils.

Furthermore, it should be emphasised that this model does not consider the wide range of socio-economic and political factors which play a crucial role in terms of their influence on the degree of soil erosion; these will be examined later. Alternative models include SLEMSA (soil loss estimator for southern Africa) and CREAMS (chemicals runoff and erosion arising from agricultural management systems).

Erosivity is a measure of the potential of the eroding agent to erode and is commonly expressed in terms of kinetic energy. The erosivity of rainfall relates to the detaching power of raindrops, which increases with increasing rainfall intensity and is largely a function of drop size. This relationship is non-linear, however, and at intensities in excess of 50 mm hr^{-1} the increase in kinetic energy is minimal as turbulence prevents further increase in drop size.

Rainfall erosivity tends to be greatest in areas where there is a marked seasonal concentration of rainfall. There are a number of rainfall erosivity indices, although one of the most widely used is the EI30 index which is a compound index of kinetic energy and the maximum 30-minute rainfall intensity. Indices of wind erosivity are based largely on the velocity and duration of the wind.

Erodibility of the soil is a measure of its resistance to detachment and transport, and depends on a number of soil characteristics, particularly texture, organic content, structure and permeability. In terms of both water and wind erosion, the most erodible soils tend to be characterised by low clay and organic contents, and poor structural stability.

Similarly, soils with an intermediate texture (fine sand to coarse loam) are more erodible than both coarse-textured sandy soils, where particle size and mass are large, and fine-textured clayey soils, which are more cohesive. The erodibility factor (K) in the Universal Soil Loss Equation can be determined from soil texture, organic content, structure and permeability.

Land use is perhaps the most significant factor influencing soil erosion, for two main reasons. First, many landuse practices leave the soil devoid of a protective vegetation cover, or with only a partial cover, for significant periods of time and second, they involve mechanical disturbance of the soil.

Specific aspects of land use often associated with accelerated soil erosion include expansion and intensification of arable cultivation, overgrazing, deforestation, certain forestry practices, site clearance in preparation for urban and industrial construction, and a number of recreational activities such as walking and skiing.

Arable cultivation has expanded and intensified dramatically in recent decades. Relatively steep slopes, formerly covered by grass or trees, have been converted to arable cropping, while an increased use of heavy agricultural machinery has resulted in compaction of the soil. This in turn has led to reduced infiltration capacity, particularly along wheel tracks, thus resulting in increased surface runoff and erosion.

Similarly, increased reliance on tillage activities, throughout the cropping cycle, has rendered soils more susceptible to erosion. This problem has been compounded by the decline in levels of soil organic matter, and hence structural stability, largely in response to increased use of inorganic fertilisers. In addition, the tendency to increase field sizes on arable land has meant that there are fewer physical breaks and barriers in the landscape, such as tree lines, hedgerows and walls, to restrict erosion.

Susceptibility to erosion is further increased if land is cultivated with the slope rather than parallel to the contours.

Overgrazing is particularly common in drought-affected parts of the developing world, such as the Sahelian region of sub-Saharan Africa, and the rangelands and communal lands of eastern and southern Africa. In a study of the impact of grazing on soils of the savanna region attribute enhanced surface runoff and erosion to compaction of the soil and destruction of the protective vegetation cover by grazing animals, and to the adoption of inappropriate burning strategies.

Deforestation, largely for logging and woodfuel purposes, is also common in many parts of the developing world. Trees are well known for their ability to protect soils from erosion, particularly on steeply sloping terrain. Their root systems, and the organic material which they supply, help to stabilise the soil, while water uptake and canopy interception serve to reduce the frequency and intensity of surface runoff.

In addition to deforestation, many forestry practices are associated with accelerated soil erosion, including the needleleaf forestry programmes which have become widespread in many areas of upland Britain. Here, erosion is most serious during the pre-planting stages of land preparation and drainage, and after harvesting. In relation to urban and industrial land use, construction and associated disturbance of land may lead to increased soil erosion. Even certain recreational activities have been implicated in this problem, including walking and skiing.

A number of socio-economic and political factors have been associated with accelerated soil erosion, particularly in the developing world. These include population pressure, skewed land resource distribution, poverty and marginalisation, increasing demand for woodfuel, innappropriate land tenure and farm policies, small size of land-holdings and poor infrastructure. In many developing countries, population growth is rapid and the demand for agricultural land and woodfuel is ever increasing.

Furthermore, agricultural systems are characterised by a skewed land resource distribution where a minority of affluent and powerful landowners control a majority of the land area. The poorest farmers are thus forced onto marginal land, which is particularly susceptible to erosion, and often end up in a vicious spiral of and woodfuel is ever increasing. Furthermore, agricultural systems are characterised by a skewed land resource distribution where a minority of affluent and powerful landowners control a majority of the land area.

The poorest farmers are thus forced onto marginal land, which is particularly susceptible to erosion, and often end up in a vicious spiral of debt. Rural-urban migration, abandonment of land and increased soil erosion are often responses to this poverty trap situation.

In many parts of the developing world, large areas of land are utilised for mono-cultivation of cash crops, which are not necessarily best suited to soil conditions, rather than for indigenous mixed food cropping. Such commercial pressure on agricultural systems, as well as contributing to the problem of marginalisation, has a detrimental effect on soil quality and is unlikely to be sustainable in the long term.

There is also little political support in terms of education, training and incentive schemes to encourage farmers to adopt more sustainable landuse practices. The establishment of appropriate and comprehensive soil conservation and land husbandry programmes is further hindered by the small size of land-holdings and the large numbers of farmers involved.

The on- and off-site effects of soil erosion are considerable. At the global scale, it is estimated that unless soil conservation measures are introduced on all cultivated land, 544 million ha of potentially productive rain-fed crop land will be lost, and agricultural production will decrease by almost 20 per cent, by the year 2000.

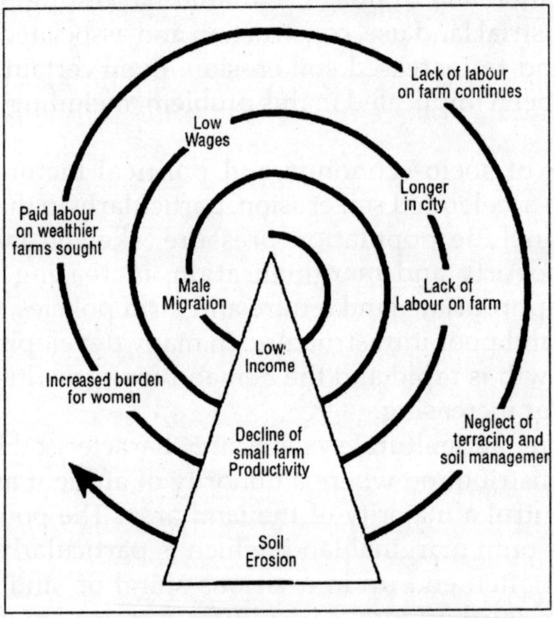

Fig. The Spiral of Poverty, Land Abandonment and Soil Erosion in the Developing World

Undoubtedly, these effects will be felt most severely in those developing countries which are least able to cope with the problem.

It should be noted that the deterioration in soil productivity is disproportionate to the amount of soil eroded, as it is the nutrient-rich and structure-supporting constituents in the topsoil which are lost most readily.

In terms of the management and remediation of soil erosion, three main strategies—agronomic, soil management and mechanical. Agronomic practices aim to protect the soil through sensible cropping programmes and are based on the encouragement of a dense vegetation cover and plant root network. Under these conditions, the time period over which the soil is left bare, and thus susceptible to erosion, is minimised.

An example of an agronomic practice is strip-cropping, in which alternate rows of different crops are arranged, usually across the slope, so that they grow and are harvested at different times, thus ensuring at least a partial cover of vegetation for much of the year. Using this approach, arable crops may be combined with strips of grass or trees. Retention of crop stubble in the soil, rather than burning, and the planting of trees or shrubs around the heads of gullies to reduce the rate of headward recession, are other examples of agronomic practices.

Soil management techniques aim to increase the resistance of soil to erosion, and focus mainly on the improvement and maintenance of soil structure. Such practices include mulching, where organic residues are added to the soil, and reduced or zero-tillage options, including direct drilling, where mechanical disturbance of the soil is minimised. Soil erodibility may also be reduced by using synthetic soil conditioners.

These consist largely of organic polymers such as PVA (polyvinyl alcohol), PAM (polyacrylamide) and PEG (polyethyleneglycol), and are applied as a surface film or are incorporated into the topsoil. Treatment is costly, however, and tends to be restricted to specialised applications such as the stabilisation of sand dune systems, and protection of the fine tilth of seedbeds.

Bio-engineering techniques, including the use of biodegradable geotextiles, have also been used to reduce the susceptibility of soils to erosion. At the Cairngorm ski slopes in Scotland, for example, these have been used successfully to facilitate the recovery of sensitive alpine plant communities.

Mechanical techniques aim to reduce the energy of the eroding agent and often involve the modification of surface topography. Examples include terracing, contour bunding and the construction of diversionary spillways to direct water away from areas which are particularly susceptible to erosion. Shelter belts may also be planted to reduce wind erosion, although these will compete with crops for soil moisture, which may be problematic in drier areas.

At the more local scale, check dams may be constructed in gully systems in an attempt to reduce erosion risk, while gabions (wire baskets filled with stones) may be installed to reduce bank erosion and headward recession. In addition to their use on agricultural land, mechanical techniques are also used to control footpath erosion. These include the installation of drainage channels to direct water flow away from damaged stretches of footpath, and the construction of elevated sections to avoid damage to poorly drained soils. In extreme circumstances, footpaths are temporarily or permanently re-routed to facilitate the recovery of damaged stretches.

In terms of the relative effectiveness of the measures, agronomic techniques tend to offer greatest protection to the soil, as they are able to control both the detachment and transport phases of soil erosion. Soil management and mechanical techniques tend to be less effective, particularly in their control of the detachment phase of erosion.

For this reason, it is usual to combine soil management and mechanical techniques with agronomic measures. Agronomic practices are often used in preference to others, not only because they are effective in the control of erosion, but also because they are relatively cheap to adopt, can be adapted to fit in with traditional cultivation methods and are acceptable to local people.

In recent decades, there has been a change in focus away from a largely technocentric approach to soil conservation, towards a broader and more ecocentric view of land management. This places emphasis not only on the soils themselves, but also on the landuse system, the socio-economic and political controls, and the wider off-site implications of soil erosion.

More important, however, is that the ecocentric approach focuses on the land users themselves and on their role in the determination of their own future. This is in marked contrast to the imposition of decisions by technical experts, which perhaps explains, at least in part, the failure of many early soil conservation projects. Specific practices which have been effective in the control of soil erosion, particularly though not exclusively in the developing world, include agroforestry, social forestry and low external input agriculture.

Compaction

Soil compaction involves the compression of a mass of soil into a smaller volume and is usually expressed in terms of dry bulk density, porosity and resistance to penetration.

The bulk density of soils which have been compacted by agricultural machinery, for example, may exceed 1.5 g cm^{-3}, while that of comparable uncompacted soils is usually between 1.0 and 1.5 g cm^{-3}. The ease with which soils are compacted depends on a number of characteristics, particularly texture, occurring most readily in soils which contain appreciable quantities of clay. The compacting force, which usually acts in a vertical direction, causes alignment of the clay platelets in a direction which is more or less parallel to the ground surface.

Alignment of the clay particles in this way often leads to the formation of compaction or cultivation pans which may be a few centimetres in thickness. Such pans tend to form at a depth of about 20-30 cm, are often characterised by a well developed platy structure, and are commonly associated with impeded drainage and restricted development of plant root networks. In silty soils, raindrop impact may lead to surface crusting which is another form of soil compaction.

Surface crusts may be several millimetres in thickness, and are commonly associated with reduced infiltration, increased surface runoff and accelerated soil erosion, and restricted germination and emergence of crops. In addition to textural controls, susceptibility to compaction also increases with increasing organic matter and water contents.

Closely related to soil compaction is the process of hardsetting which involves an increase in bulk density but without the application of an external load. Hardsetting soils are characterised by their low structural stability. During and after wetting, slaking and collapse of aggregates lead to uniaxial shrinkage and dispersal of clay and silt and, on drying, the soils harden without restructuring.

Acceptance of hardsetting soils as a distinctive group has been hindered by the difficulties experienced in distinguishing hardsetting characteristics from other forms of soil behaviour.

One of the main causes of soil compaction is the excessive use of heavy agricultural machinery, and compaction is particularly severe along wheelings left by these vehicles. For example, bulk density values of 2.2 g cm^{-3} and 1.3 g cm^{-3} have been recorded in soils of wheeled and inter-wheel areas respectively. The intensification of arable cultivation in recent decades has been facilitated to a large extent by increased mechanisation.

Most procedures in the cropping cycle, from tillage and seedbed preparation, through drilling, weeding and agrochemical applications to harvesting, are now largely mechanised, particularly in the developed world. The timing of cultivation in relation to precipitation and soil water levels is known to be critical with respect to soil compaction. If soils are cultivated when they are near or above field capacity, or their plastic limit, then severe structural damage and compaction are likely to occur.

Soil compaction is not restricted to arable land but is also common on land used for grazing and forestry. Grazing animals are well known for their ability to cause compaction, or poaching, especially during wet conditions.

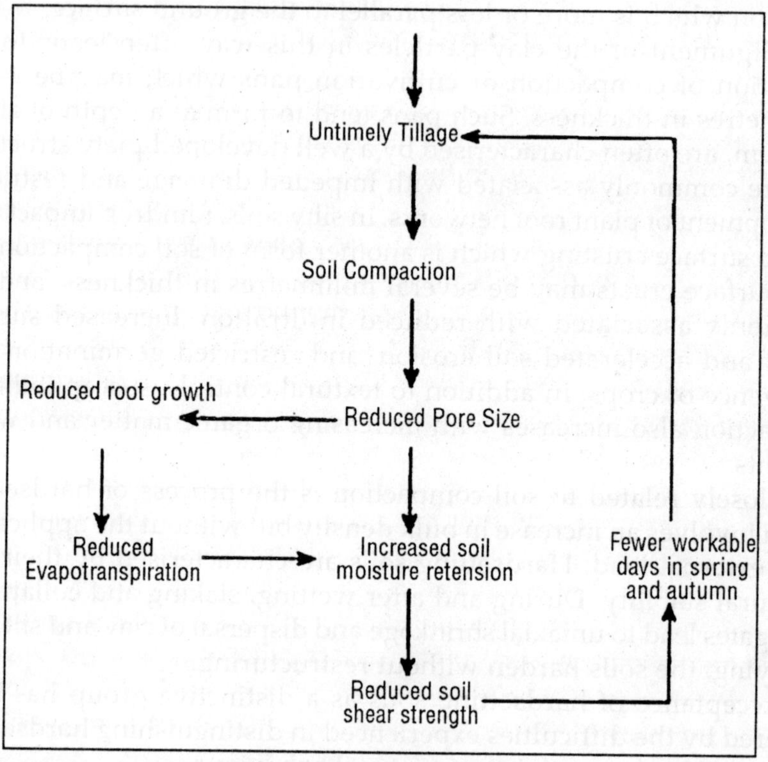

Fig. Feedback Effects of Soil Compaction Due to Untimely Tillage

It is particularly common to find poorly drained or puddled areas in the vicinity of food troughs or gateways where animals tend to congregate. In a study of the impact of grazing animals on soils of the savanna region, it was reported significant compaction and loss of protective vegetation cover in grazed plots.

In forestry plantations in upland Britain, heavy agricultural machinery is often used during the preparation and drainage of land prior to planting, and during harvesting operations; such mechanisation is often associated with the disturbance and compaction of soils. Similarly, soil compaction has been reported in association with the clearance of tropical rainforest, particularly where mechanical methods are used in preference to the traditional slash and burn approach.

Soil restoration following mineral extraction, quarrying and opencast coal mining has also been implicated in the soil compaction problem. Compacted horizons result largely from the passage of heavy machinery during soil stripping and replacement, and from the shear forces produced during the lifting process. In clay and loam topsoils which had been stockpiled, it was reported bulk density values that were 20 per cent higher than in similar topsoils at undisturbed control sites nearby.

Elevated values are not restricted to topsoils, however. It was observed dramatic increases in bulk density at the top of the subsoil, at about 0.3 m in depth, in restored soils in northeast England; values of 1.64 g cm^{-3} and 1.46 g cm^{-3} were recorded in the restored and unmined control sites respectively.

Compaction has adverse effects on a number of soil characteristics. Increased bulk density and the consequent decrease in porosity are associated with both increased waterlogging and poor aeration; these changes in turn have a detrimental effect on the thermal characteristics of soils.

Under these conditions, plant growth may be restricted, particularly during the early stages of germination and emergence, and during the main phase of root network development. Similarly, the increased incidence of root rot in soils with compacted cultivation pans has been widely documented. Other characteristics that may be adversely affected by compaction include the size and diversity of soil organism populations, and the incidence of certain crop pests and diseases may also increase.

Soil compaction has been associated with declining crop yields in many areas. Expert present yield data for a number of crops grown in soils with a variety of textures, ranging from sand to clay. Yields from compacted sites were found to be 25-74 per cent lower than those from uncompacted control sites. Similarly, yield reductions of up to 50 per cent were observed for corn on compacted clay soils in Quebec.

The financial impact of soil compaction is difficult to assess due to the influence of a number of interrelated factors. In addition to yield reductions, damaged soil structure may have to be restored, surface runoff and soil erosion may increase, fertiliser usage becomes less efficient due to increased leaching losses, water loss by evaporation may increase, together with the operational costs of irrigation, and in soils which are susceptible to waterlogging, expensive drainage systems may have to be installed. It is estimated that the cost of fuel required for tillage of compacted soils may be up to 35 per cent greater than that for uncompacted soils.

Successful management and remediation of soil compaction centres on two main approaches—improvement and maintenance of soil structure, and appropriate tillage practices. Good soil structure provides the basis for increased mechanical strength and stability, improved water retention and aeration, and protection of plant nutrients from leaching. Soil structure can be improved and maintained through the promotion of sensible crop rotation practices, particularly those which include significant periods of ley pasture, mulching and stubble retention.

Unlike continuous cropping, such practices encourage the accumulation of soil organic matter and the development of strong plant root networks, both of which are essential ingredients in the development of good soil structure. Soil structure also benefits from the addition of certain nutrients, particularly the multivalent base cations which play a crucial role in the aggregation process.

In addition to soil structural improvements, good tillage practices are an essential component of effective management and remediation of soil compaction.

Tillage of agricultural land is practised for a number of reasons—first, to create a fine tilth which facilitates effective germination and emergence of seedlings, second, to remove weeds which compete with crops for nutrients, water and light, third, to incorporate crop residues into the soil with the aim of improving soil structure and nutrient content, and fourth, to improve drainage and aeration in the root zone (rhizosphere).

Although tillage is therefore of benefit to the soil, when mechanised it can lead to soil compaction. The timing of tillage is particularly important in this respect, and it should be avoided in wet conditions, particularly if soil moisture content is likely to exceed the field capacity or plastic limit.

In an attempt to alleviate compaction resulting from the excessive use of heavy agricultural machinery, reduced or zero-tillage practices have been widely adopted.

Such practices are inappropriate for poorly drained soils, however, where conventional tillage helps to improve drainage, aeration and the thermal characteristics of the topsoil. They should also be avoided on excessively drained soils, where increased nitrate leaching and denitrification may lead to nitrogen deficiency, and on steep slopes where seed slots may be subjected to increased erosion.

Thus, it appears that the benefits of reduced and zero-tillage are felt most strongly on medium-textured soils which are well drained and which possess a well developed structure. In some circumstances, compaction and associated poor drainage cannot be improved through good structural management and appropriate tillage practices alone, and installation of artificial drainage may be necessary.

Water Excess and Deficit

Excess water may be present in soils for a number of reasons. Compaction, for example, in the form of surface crusting, subsurface cultivation pans and hardsetting, often leads to reduced infiltration and restricted permeability. Excess water is also common in heavily textured clay soils, particularly in high rainfall areas, and in areas which are flat, low lying and prone to flooding. If excess water is found within about 40-50 cm of the surface, then the adverse effects on soil aeration and temperature are likely to lead to restricted root growth and crop performance.

In many cases, excess water may be removed from the soil through both sensible management of soil structure and appropriate tillage practices. Frequently, however, installation of artificial drainage is necessary. The aim of any artificial drainage programme is to lower the water table so that the amount of plant-available water in the rhizosphere is maximised.

If the water table is lowered too much, the soil may become drought-susceptible. Conversely, if the water table remains too high, waterlogging and poor aeration are likely to continue. The efficiency of drainage depends largely on the lateral spacing, depth and size of drains. Generally, efficiency improves with decreased lateral spacing, although at spacings of greater than around 6 m the rate of inflow to the drains is independent of spacing.

Drain depth controls the height of the water table and usually occurs at about 50 cm to 2 m. In terms of size, the ability of a drain to carry water increases with increasing cross-sectional area. The choice of drainage system depends on a number of factors, the most important being economic controls (installation and maintenance costs, expected yield improvements and availability of subsidies), management considerations (crop and tillage requirements), soil characteristics (texture, hydraulic conductivity and type of drainage problem) and climatic controls (rainfall amount, intensity and seasonality).

There are a number of specific drainage techniques which can be employed, either singly or in combination, depending on the type and extent of the drainage problem. Those most commonly used include subsoiling, mole drainage, pipe or tile drainage, open ditches or dykes, and regional or arterial drainage. Subsoiling is a form of deep ploughing and is often used to break up compacted cultivation pans and indurated layers within the soil.

The critical depth for this practice is about 20-50 cm; if it is shallower, then lifting of the soil may occur and if deeper, consolidation of the soil may result. Subsoiling is most effective when the soil is relatively dry and heavy; in wet conditions, plastic deformation and smearing may occur during passage of the subsoiling implement, and this is likely to have an adverse effect on soil structure.

Mole drainage is produced by drawing a mole plough, which is a bullet-shaped instrument usually less than 20 cm in diameter, through the soil to produce a continuous passage. Mole drains are established at a depth of about 40-60 cm, with a lateral spacing of about 2-5 m. They are not permanent features and are susceptible to collapse and sediment blockage.

Consequently, they have to be renewed on a regular basis, usually after a period of about 5-7 years. During the installation of mole drainage, the moisture status of the soil is of crucial importance. If conditions are too wet, smearing and consolidation may result, while shattering may occur if the soil is too dry. In spite of the temporary nature of mole drainage, it is often adopted in preference to other drainage practices because it is cheap and easy to install. It is also commonly used as a form of secondary drainage in combination with primary tile or ditch drains.

Pipe or tile drainage consists of plastic or ceramic pipes of about 10-30 cm in diameter, and is installed at a depth of 0.5-2.0 m. The pipes usually contain perforations or slits in order to speed up the transfer of water from the soil. Once the pipes have been installed, the ditches are often back-filled with gravel, polystyrene or fuel-ash. This helps to improve the flow of water to the drains, while acting as a filter to remove fine sediments which may otherwise lead to drain blockage.

Open ditches or dykes are usually constructed at field margins and are relatively cheap and easy to maintain. This type of drainage has been used since settled agriculture began, and in Britain dates back to the Roman period. In some circumstances, drainage may need to be improved at the regional scale by arterial drainage. This can be achieved in a number of ways, including straightening of river channels to increase gradient and improve flow, deepening of river channels by dredging, and pumping of water from agricultural land.

Flood prevention is also an important aspect of arterial drainage. In coastal areas, this is achieved through the construction of barrier systems which form part of sea defence strategies. Similarly, in low-lying floodplain areas, it may involve the construction of raised river banks or levées.

One of the best-known examples of this type of large-scale reclamation is the Polder scheme in the Netherlands, where large areas were reclaimed from the sea in the low-lying coastal fringes. Since the Zuyder Zee (Reclamation) Act of 1918, four polders have been reclaimed, creating about 165,000 ha of new land. This involved the construction of large dykes which were fed by lateral drains spaced at 1.6-2.0 km intervals. These drains were supplemented by main drains spaced at 300 m intervals and then a coarse network of field drains spaced at 8-24 m intervals. Although such large-scale drainage programmes have produced some of the world's most fertile and productive land, through improvements to a range of soil physical and chemical characteristics, they have also created problems.

Perhaps the most notable of these is soil shrinkage, which is particularly common in areas dominated by marine clays and peats. In parts of the Fens of eastern England, for example, it is estimated that the surface has been lowered by as much as 5 m following the long history of reclamation and drainage. In an area which is already low lying, surface lowering places even greater pressure on existing flood prevention and drainage schemes, many of which need to be upgraded. Also of concern in areas where large-scale reclamation and drainage are widespread, is the destruction of important wetland habitats.

Soil water deficit occurs most commonly in areas where rainfall is low in comparison with potential evapotranspiration. It is also common in coarse-textured, sandy soils which are excessively drained and low in organic matter, and in fine-textured, clay soils which contain large vertical cracks or which are poorly structured and therefore of low permeability; such soils have a low plant-available water capacity. Soil water deficit does not necessarily occur throughout the year but is most common during warmer and drier periods.

In Britain, for example, it rarely occurs in the October-March period but is common in the April-September period, especially in southern and eastern areas where it may occur in three to five years out of ten. In the USA it is particularly common in the central and southwestern states, but is relatively rare in the southern and eastern states where rainfall is greater and more reliable. Soil water deficit has the greatest impact where it occurs during the main period of crop growth, when it is demonstrated by excessive wilting of plants. In some circumstances, intense evaporation may lead to salinisation and sodification of soils.

The main way in which soil water deficit is managed is by irrigation. This has been practised since the beginning of settled agriculture, not only in areas where rainfall is low throughout the year, but also where seasonal drought is a problem, in spite of respectable annual rainfall totals. It is used to replenish the soil moisture store and to leach toxic salts out of the rhizosphere, and as such it is a fundamental component of agriculture in many dryland areas. The effectiveness of irrigation depends on the method and timing of water application, and on the amount and quality of water applied. There are three main methods—surface application, overhead application and subsurface application.

Surface application usually involves flood and furrow irrigation where water is diverted from existing river channels or canals directly onto the ridge and furrow systems of agricultural fields. This low cost and low technology approach is often preferred in developing countries, although it is also widely used in the developed world. One of the main problems with this method, however, is that capillary action may lead to redistribution of salts from the wet furrow areas to the drier ridges where the crops are often planted.

Overhead irrigation methods include spray or sprinkler, and trickle or drip systems. Spray irrigation systems involve the high pressure application of water through upright nozzles which may be attached to fixed or moveable pipes in a variety of ways, such as self-propelled water guns and centre-pivot systems. Although water can be applied uniformly and efficiently using these methods, equipment costs may be prohibitive and evaporative losses are high, particularly on windy days. Moreover, soil erosion may occur as a result of overhead application, particularly in the early stages of crop growth when vegetation cover is low. Trickle irrigation involves the application of water through flow regulating emitters which are placed at regular intervals in close proximity to the plants. In this way water is supplied more directly to the plant and evaporative losses are minimised.

Perhaps the most effective methods of irrigation involve subsurface applications of water. Here, ditches and underground drains or pipes are used and, although their installation can be costly, evaporation, soil structural damage and erosion losses are minimised.

In addition to irrigation, soil water deficit may also be managed through the control of evaporation from the soil surface. This is most commonly achieved through the addition of organic mulches, or through the establishment of a vegetation cover which can offer some degree of shade. These strategies are usually combined with sensible irrigation practices in order to increase water use efficiency in dry environments.

CHEMICAL PROBLEMS

Acidification

Acidification has a number of natural and anthropogenic causes.

The main natural causes are long term leaching and microbial respiration. The acids found in rainwater (carbonic acid) and in decomposing organic material (humic and fulvic acids) can stimulate leaching by dissociating into H^+ ions and their component anions which then displace or attract base cations from the soil exchange complex.

Leaching of bases is most common where precipitation exceeds evapotranspiration, and includes many eastern parts of North America, and northwestern parts of Europe, where soils have been subjected to leaching throughout most of the 10,000-15,000 years of post-glacial time.

Microbial respiration also leads to soil acidification through the production of CO_2 (carbon dioxide), which is dissolved in soil water to form carbonic acid. Other natural processes associated with soil acidification are plant growth and nitrification. During plant growth, nutrient base cations are obtained through root systems in exchange for H^+ ions, thus leading to increased soil acidity. Nitrification is an oxidative process of organic decomposition whereby NH_4^+ (ammonium) ions are converted to $NO_3^{''}$ (nitrate) ions by nitrifying bacteria, with H^+ ions as a by-product:

$$NH_4^+ + 1.5O_2 \Rightarrow NO_3^- + 4H^+$$

These ions are then available for displacing and attracting base cations from the soil exchange complex, thus leading to soil acidification.

The main anthropogenic causes of acidification include certain landuse practices, such as needleleaf afforestation, excessive use of inorganic nitrogen fertilisers, land drainage, and acid deposition resulting from urban and industrial pollution. Needleleaf afforestation has been associated with the acidification of soils and surface waters for a number of reasons. First, needleleaf trees produce litter which is very acidic in comparison with most broadleaf species. Second, because of their high canopy surface area, needleleaf trees are able to 'scavenge' acid pollutants from the atmosphere, later releasing them into the soil via throughfall and stemflow.

Third, due to modifications of the surface and soil hydrology by drainage channels and shallow root networks, water transfer is rapid and is concentrated either at the surface or in the uppermost layers of the soil. Under these conditions, the residence time of the water in the soil is limited, as is the depth to which it can percolate. Thus, the contributions of weathering and ion exchange reactions to the buffering process are limited.

Excessive use of inorganic nitrogen fertilisers in agricultural systems has also been associated with soil acidification, partly through the process of nitrification. If levels of NO_3^- ions in the soil are in excess of plant requirements, they will behave as mobile anions, thus encouraging the leaching process. The acidifying effect of nitrogen fertiliser was demonstrated in a four-year experiment during which different loadings of NH_4NO_3 (ammonium nitrate) fertiliser were applied to grassland plots; soil pH declined markedly with increased fertiliser loading.

Land drainage is another important cause of soil acidification, particularly in soils which contain appreciable quantities of sulphide minerals. Once the land has been drained, soil aeration improves and sulphide minerals are oxidised to form sulphate compounds. Sulphate ions can then combine with H^+ ions in the soil to produce H_2SO_4 (sulphuric acid). Under extreme circumstances, soil acidity may fall below pH 3.0 as in the case of the acid sulphate 'cat-clay' soils which are common in mangrove swamp environments.

Another important anthropogenic source of acidity in soils and surface waters is atmospheric deposition. Here, gases derived from industrial and motor vehicle emissions, particularly SO_2 (sulphur dioxide) and NOx (oxides of nitrogen), are either dissolved in precipitation *(wet deposition)* or deposited directly *(dry deposition)*. Essentially, acidity is derived from H_2SO_4 (sulphuric acid) and HNO_3 (nitric acid) which undergo dissociation in rain and soil water.

Values of pH for acid deposition in industrial areas of Europe and North America are often less than 4.0. Some of the lowest values, however, (pH < 3.0) have been observed in acid mists and fogs, a phenomenon known as *occult deposition*. This type of acidity is common in upland areas of Britain where it may be scavenged by the canopies of needleleaf forestry plantations and then transferred to the soil.

The acidification of soils has an impact on a number of soil characteristics. It was report pH reductions of up to 1.5 units over a 55-year period in southern Sweden; these reductions were attributed to high levels of acid deposition derived from other areas of northern and western Europe. Nutrient loss through leaching is another effect of soil acidification.

Expert suggests that the supply of available calcium and magnesium in topsoils will be halved within twenty to thirty years, due to continued acidification and leaching. In addition to nutrient losses, the mobility of aluminium in soils increases as they acidify, especially when soil pH falls below about 5.0. Levels of aluminium are often particularly high in the B (spodic) horizons of podzolic soils.

If organic acids predominate in the soil, aluminium is mobilised in the form of soluble organo-metallic complexes. If mineral acids predominate, however, then aluminium is mobilised in its ionic, labile-monomeric form (Al^{3+}); this form of aluminium is particularly toxic to many freshwater organisms, including fish. In a similar way to aluminium, heavy metals, such as lead, zinc and cadmium, are more readily mobilised in acidic soils.

Acidification of soils, and associated nutrient leaching, has also been implicated in damage to trees in forested areas, particularly in central Europe and Scandinavia. It is more likely, however, that nutrient leaching acts synergistically with other factors including ozone pollution, acid deposition, NH_4^+ (ammonium) uptake, drought and frost to produce stress in the tree.

Signs of tree damage include needle discoloration, crown defoliation and deformed branch structures. Increased soil acidity and associated changes in aluminium mobility also have serious implications for the quality of surface waters which receive drainage from acidified soils.

Soils vary dramatically in their ability to buffer acidity, a characteristic known as the *buffering* or *acid neutralising capacity*. The buffering capacity of a soil is defined as the amount of acid that needs to be added to cause a reduction in pH of one unit. Soils which contain significant quantities of base-rich, weatherable minerals have a high buffering capacity, whereas those which are dominated by quartz and similarly resistant minerals have a low buffering capacity.

In Britain, for example, soils with the highest buffering capacities are found in eastern and southern areas where calcareous parent materials are particularly common. Soils with the lowest buffering capacities are found in areas with base-poor, often siliceous, parent materials which predominate in upland areas in the west and north.

Surface water acidification is widespread in areas where the geology and soils are base-poor, and levels of acid deposition are high. The hydrological pathway taken by drainage waters and their residence time in the soil are particularly important factors in this respect. Water flowing rapidly over the surface or through the uppermost soil horizons undergoes little buffering and may carry organic acids from the soil, therefore contributing significantly to surface water acidification.

If water is allowed to percolate slowly through the deeper mineral horizons, however, its acidifying effect is likely to be limited due to increased buffering. Surface water acidification is exacerbated in areas where needleleaf afforestation is widespread, as is the case in many upland parts of northern and western Britain. Experts examined the chemistry of streams under a needleleaf forest plantation and adjacent moorland in North Wales. The streams under forest were more acidic, and contained higher levels of aluminium, SO_4^{2-} (sulphate) and Cl^- (chloride). Similarly, in a study of water quality in streams draining land used for rough grazing and forestry, Experts found levels of ionic aluminium to be particularly high in the forest streams. These findings are also consistent with observations in many forested areas. Elevated levels of aluminium and heavy metals in potable water supplies have been implicated in certain mental disorders in humans, notably Alzheimer's disease, although the medical evidence is not conclusive.

In addition to the effects of land use, stream flow conditions have an important influence on surface water acidity. Storm runoff is known to be more acidic than base-flow because it flows rapidly over the surface of the soil, or through the uppermost horizons, and is unable to percolate into the deeper mineral horizons where it might be buffered. Experts report pH values of 4.4 and 6.4 for storm-flow and base-flow respectively in central Wales. Similarly, acid surges associated with storm runoff and snowmelt have been reported in eastern Scotland.

There are a number of approaches to the management and remediation of soil and surface water acidity, including liming, sensible forestry management and reduction of acid emissions into the atmosphere. Liming has long been practised on agricultural land to remedy the problems of acidity, slow organic matter turnover, poor nodulation in some legumes, calcium and molybdenum deficiency, and aluminium and manganese toxicity.

Liming materials most commonly used include ground limestone, chalk, marl and basic slag, the main active constituent being $CaCO_3$ (calcium carbonate); other minor components include quicklime (CaO), slaked lime ($Ca(OH)_2$) and $MgCO_3$ (magnesium carbonate). The lime requirement of a soil varies depending on its buffering capacity and is usually expressed as the amount of $CaCO_3$ (t ha^{-1}) required to raise the pH of the top 15 cm of soil to the desired value.

In temperate areas, the ideal soil pH is about 6.5 for arable crops and about 6.0 for grassland. In tropical areas, however, pH values of about 5.5 are often preferred, particularly in soils with high exchangeable aluminium contents, where phosphorus availability may be restricted under more alkaline conditions.

Liming is also used to combat surface water acidity. In many parts of Scandinavia, for example, particularly in Sweden, lakes are often limed directly. This practice is rather costly, however, as the time period over which the lime is effective *(hydraulic retention time)* is relatively short, mostly ranging from several weeks to two years. Furthermore, fish populations do not benefit fully from such treatments as they tend to breed and spawn in the feeder streams which are not limed. In an attempt to overcome these problems, catchment liming is sometimes practised.

Here, the whole catchment benefits from treatment, and the time-scale over which the treatment is effective is usually much greater than when only the lakes are treated. The cost of treatment may be reduced substantially by strategic applications of lime to specific sites within the catchment, such as spring and seepage areas, and stream headwaters. In some catchments, flow-activated lime wells have been used to supply lime to streams during periods of high flow when acidification is known to be a major problem.

Acidification of surface waters in afforested areas may be controlled through sensible forestry management practices, rather than by chemical means. For example, the termination of drainage ditches before they enter water courses, sometimes in deep check drains which run across the slope, will help to increase the residence time of water in the soil, thus facilitating the buffering process. Other useful practices include not planting in the riparian buffer zones adjacent to water courses, and increasing the proportion of broadleaf trees in plantations.

Most of the acidity mitigation strategies are curative in nature rather than preventative. In the long term, perhaps the most effective strategy is to reduce acid emissions, particularly of SO_2 (sulphur dioxide), into the atmosphere. European emissions of SO_2 from coal- and oil-fired power stations have decreased dramatically since 1980, partly as a result of installing flue-gas desulphurisation equipment and partly by switching to other energy sources such as gas, nuclear power or wind power.

During the mid-1980s, in an attempt to form a more objective and scientific basis to the issue of cuts, and how large they need to be so that long-term ecosystem recovery is ensured, the 'critical loads' approach was devised. The critical load is the highest loading of acid that can be taken by an ecosystem before long-term damage occurs. In some of the most sensitive areas of Britain, it was found that the reduction required for ecosystems to recover is about 90 per cent.

This is considerably greater than the 60 per cent reduction, by the year 2005, to which Britain is currently committed.

Salinisation and Sodification

Salinisation describes the accumulation of salts in the soil, whereas sodification or alkalisation refers to the dominance of the soil exchange complex by Na^+ ions. These processes occur largely in semi-arid and arid environments where evaporation exceeds precipitation, and where soil parent materials and ground waters are rich in sodium salts.

In such dry environments, soil water movement is driven largely by evaporation and occurs in an upward direction by capillary action. As water evaporates, salts precipitate out to form saline (solonchak or white alkali) soils. Slight dissolution of these salts and the consequent release of their constituent ions into solution may lead to dominance of the soil exchange complex by Na^+ ions, resulting in the formation of sodic (solonetz) soils.

Saline soils occur most commonly in low-lying floodplain areas where the water table is relatively high, thus leading to intense capillary action. In contrast, sodic soils are found most commonly on slopes immediately above valley floors and flood plains where the water table is lower and drainage potential is greater.

On higher slopes in such areas, a leached and more acidic variety of the solonetz soil, known as a solod, is often found. This catenary association of solonchak, solonetz and solod soil types is widely observed in semi-arid areas, including the Indus valley of Pakistan, the Nile valley of Egypt and Sudan, and the Tigris and Euphrates valleys in Iraq and Syria.

They are also found in many parts of central and southern Australia and around the Caspian and Aral seas in central Asia. Even in England and Wales, where the climate is relatively wet, salt-affected soils occupy about 6 per cent of the agricultural land area, being especially common in low-lying coastal and estuarine areas.

Soil salinity is measured in terms of the electrical conductivity of a saturated extract (ECe), whereas soil sodicity is established from the exchangeable sodium percentage (ESP) or sodium adsorption ratio (SAR). Soils are classified as saline, if the ECe value is 4.0 mmho cm^{-1} or more, pH is 8.5 or less and ESP is less than 15 per cent.

Sodic soils are non-saline, have a pH of more than 8.5 (usually less than 10.0) and an ESP value of 15 per cent or above. Soils are classified as saline-sodic if they are both saline and have an ESP value of 15 per cent or above.

The salinity of ground waters and other water supplies, which may be used for irrigation purposes, is also assessed on the basis of electrical conductivity (EC). Most water supplies have EC values of 0.15-1.5 mmho cm^{-1}, and are said to be of high salinity if the EC value exceeds 0.75 mmho cm^{-}.

Although salinisation and sodification occur naturally in semi-arid and arid environments, they are often exacerbated as a result of human activity. In parts of southwest Australia, for example, removal of indigenous eucalyptus forest has resulted in extensive salinisation and sodification of soils. This has occurred because the deeply rooted trees have been replaced by shallow-rooted grasses and crops, which are less effective at lowering the ground-water level.

Capillary action is most intense, and salinity and sodicity are greatest, in soils where the water table is within about 2 m of the surface. Another important cause of soil salinity and sodicity is poor irrigation practice. Overwatering leads to a rise in the water table which in turn causes enhanced capillary action. Similarly, poor maintenance of irrigation channels and canals results in leakage of water onto adjacent agricultural land. This has contributed to increased soil salinity and sodicity in parts of the Indus valley in Pakistan, for example.

In the Nile valley in Egypt, river impoundment resulting from construction of the Aswan dam has been associated with a deterioration in soil quality in what were once very fertile floodplain areas. Prior to this, flooding would occur annually in response to heavy seasonal rains in the mountainous source regions of the river.

The seasonal flood waters were particularly important to agriculture in the Nile valley for three main reasons—they were an important source of irrigation water, they brought with them fertile silt deposits which were incorporated into the soils, and they helped to flush out salts from the soil.

Since river impoundment, the benefits of seasonal flooding have been removed, and the resulting decline in soil quality has been compounded by agricultural intensification, in order to meet both the needs of a rapidly growing population and the demands of overseas markets for cash crops.

Salinisation and sodification of soils are not only associated with arable land, but may also occur in pastoral landuse systems. Experts examined the salt regimes of soils under grazed and enclosed (ungrazed) natural grassland in the Flooding Pampa of Argentina. Here, soils are affected to varying degrees by salts and Na^+ ions, the water table is high and flooding is common during the wet season.

The salt concentration in the topsoil of the grazed land increased dramatically and episodically after flooding, whereas on the ungrazed land it did not. This difference was attributed to compaction and impeded drainage, and to reduced vegetation cover on the grazed land, which led to increased evaporation and enhanced capillary action in the soils.

Soil salinity and sodicity have a detrimental effect on both chemical and physical aspects of soil quality. In a chemical context, high salinity and sodicity are associated with elevated soil pH, and under these conditions the availability of certain plant nutrients is reduced, resulting in severe disturbance of the plant nutrient balance as a whole.

Elevated salt and Na^+ concentrations in soils are also highly toxic to many plants, although tolerance levels vary between different species. Crops such as barley, cotton and sugar beet, for example, have a relatively high tolerance level, whereas sugar cane, onion and lettuce have a very low tolerance level. As with arable crops, grasses vary in their tolerance to soil salinity and sodicity. Bermuda grass, for example, has a high tolerance level whereas Guinea grass is much less tolerant.

Of particular significance in sodic and saline-sodic soils is the destabilisation and breakdown of soil structure. Deterioration of soil structure is not usually a problem in saline soils, provided that the concentration of soluble salts remains high. However, if salts are depleted by leaching, and Na^+ ions begin to dominate the soil exchange complex, then swelling and deflocculation of clays may occur, thus resulting in aggregate dispersion. These changes, together with translocation of the dispersed clay, may lead to reduced macroporosity and permeability. Although clay swelling may be reversed through increases in salt concentration, deflocculation and translocation are not reversible and may cause permanent deterioration of soil structure.

Soil salinity and sodicity management and remediation are complex issues, and may involve a number of approaches. Appropriate irrigation and drainage techniques are critical in the effective management of saline, saline-sodic and sodic soils. In addition to the amount of water applied, the quality of the irrigation water is important. Use of water which contains high concentrations of salts can lead to increased salinity and sodicity in soils.

As the water evaporates, ionic concentrations increase and ions in solution (particularly Na^+) are exchanged with those held on the soil exchange complex (notably Ca^{2+}) until a state of equilibrium is achieved. Saline-sodic soils in particular require careful management in this respect because, although structural stability is maintained if the concentration of salts in the soil solution is high, if soluble salts are depleted by leaching, structural deterioration will occur as Na^+ ions begin to dominate the soil exchange complex.

Reclamation of such soils requires leaching with water of a low enough SAR to facilitate Ca^{2+} exchange for Na^+, but a sufficiently high total salt concentration to maintain soil structure and permeability. As the soil ESP is gradually reduced, the salinity of the leaching water may also be reduced until reclamation is complete.

Intermittent leaching is more effective than continuous ponding as it allows time for salts to diffuse into the depleted outer regions of aggregates from where they are removed during the next leaching phase. During continuous ponding, water flows preferentially through the larger inter-aggregate channels and does not remove salts trapped within the aggregates.

Soil salinity and sodicity can also be alleviated using chemical amendments. Sodic soils, for example, can be improved by treatment with gypsum ($CaSO_4.2H_2O$). About 3 tonnes of gypsum are required to reduce by one unit the ESP of a volume of soil which is 1 ha by 15 cm deep. During treatment, Ca^{2+} ions displace Na^+ ions from the soil exchange complex. This benefits the soil in a number of ways—nutrient status is improved, the degree of alkalinity is reduced and structural stability is increased. If $CaCO_3$ (calcium carbonate) is present in the soil, Ca^{2+} ions may be released through the addition of sulphur to the soil. The acidity generated during the oxidation of sulphur to SO_4^{2-} (sulphate) leads to dissolution of the $CaCO_3$; H_2SO_4 (sulphuric acid) may also be used for this purpose.

In addition, salinisation and sodification of soils may be controlled by manipulation of surface microrelief and by reducing evaporation from the soil surface; this helps to reduce the intensity of capillary action. On land which has been irrigated using the flood and furrow method, capillary action may lead to redistribution of salts from the wet furrow areas to the drier ridges where the crops are often planted; to some extent, this problem can be alleviated by modifying the shape of the ridges. Salt concentrations tend to be particularly high on the ridge tops, so planting in this position should be avoided.

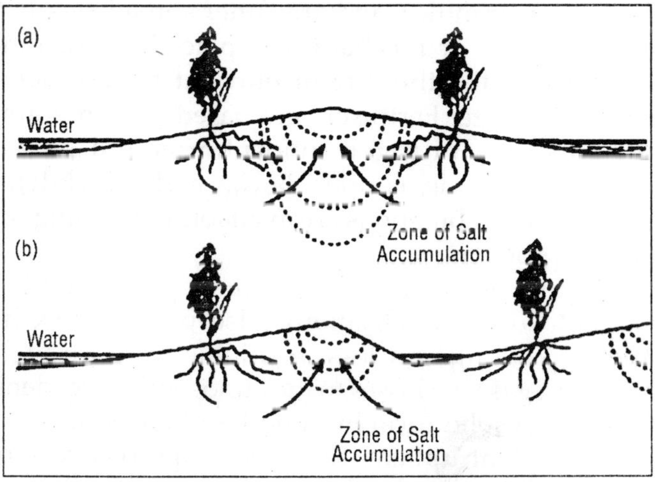

Fig. Microtopography of Ridge and Furrow Systems Designed to Avoid Salinity Damage to Crops: (a) Paired Crop Rows on Broadly Sloping Ridges, (b) Single Crop Rows on Asymmetric Ridges

Instead, planting should take place on the sides of ridges where salt concentrations are relatively low. Evaporation may be reduced through the addition of organic mulches to the soil surface. Mulching plays a crucial role in the conservation of soil moisture in soils of both warm and dry environments.

If levels of soil salinity and sodicity are not too high, it may be possible to utilise salt-tolerant crops and grasses with minimal reductions in yield, thus avoiding the need for ameliorative treatments. The replanting of trees can also reduce the level of the water table and thus the extent of capillary rise of salts towards the surface.

Agrochemical Pollution

In recent decades, the use of inorganic fertilisers has increased dramatically at the expense of more traditional organic nutrient treatments. Between 1950 and 1985, the global use of fertilisers increased from 14 million tonnes to 125 million tonnes, an increase of almost 900 per cent.

Inorganic fertilisers are used in preference to organic treatments because the nutrients are in a more readily available form and are released rapidly after application. Organic material releases its nutrients slowly, through decomposition processes, and only when conditions are suitable (warm and moist), not necessarily when crops need them.

Fertilisers are applied in a variety of forms—solution, suspension, emulsion and solid. The solid forms vary in particle size from fine powders to coarse granules, and are either spread evenly (broadcast) over the soil surface or mechanically placed, by drilling, into the rhizosphere; generally, the rate of nutrient release decreases with increasing particle size. Fertilisers are based on compounds of plant macronutrients (e.g. nitrogen, phosphorus and potassium) and micronutrients (e.g. zinc, copper, boron and molybdenum), and a variety of nutrient combinations are available depending on the nature of the nutrient problem.

There are five main fates of fertilisers applied to the soil—plant and animal uptake (immobilisation), adsorption and exchange in the soil (fixation), leaching and loss in soluble form through drainage, volatilisation and gaseous losses to the atmosphere (e.g. denitrification), and surface loss in solid form by runoff and erosion.

In terms of plant uptake, the percentage recovery of nutrients varies markedly between the different types of fertiliser. During the first year of application, the recovery of nitrogen from inorganic nitrogen fertilisers is about 50-65 per cent, whereas from organic manures it is only about 20-30 per cent. Similarly, the recovery of phosphorus and potassium from inorganic fertilisers is about 5-15 per cent and 75 per cent respectively. In terms of fixation in the soil, many of the phosphorus and potassium fertilisers are of relatively low solubility and the nutrients released are often strongly adsorbed. In contrast, many of the nitrogen fertilisers are highly soluble, nutrient release is rapid, and in most soils adsorption is limited.

The differences in degree of immobilisation by plants and animals, and in the extent of fixation in the soil, help to explain why leaching losses of nitrogen are usually far greater than those of phosphorus and potassium. Levels of nitrogen in drainage waters, for example, are often in the range 15-150 kg ha^{-1} a^{-1}, whereas levels of phosphorus are generally <1 kg ha^{-1} a^{-1}; even in exceptional circumstances in eroded areas, phosphorus levels rarely exceed 10 kg ha^{-1} a^{-1}.

Nitrogen losses of up to 30-35 per cent of that applied have been recorded for both leaching and denitrification. Leaching is most common in coarse-textured and well drained soils, whereas denitrification losses are greatest in fine-textured, waterlogged and poorly aerated soils.

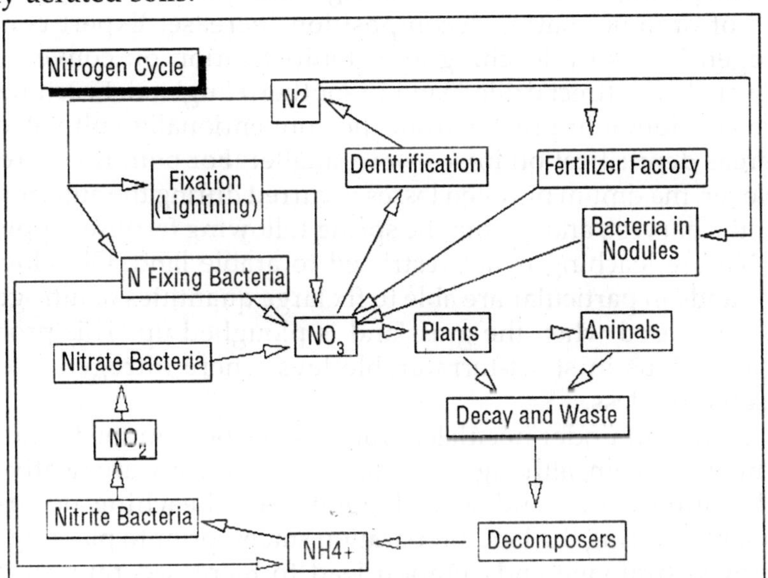

Fig. The Soil Nitrogen Cycle and the Fate of Nitrogen Fertilisers Applied to Soils

Similarly, surface losses are greatest at sites which are most susceptible to surface runoff and erosion. In terms of the environmental problems associated with fertiliser use, perhaps the area which has received most attention in recent years, by way of research, is that of nitrate leaching. In the last few decades, levels of nitrate in water supplies have increased dramatically, particularly in intensively cultivated areas where inputs of nitrogen fertiliser have been high. This association is not necessarily causal, however, and may be indirect. Nitrogen fertilisers are not the only source of leached nitrate. It was found that here much of the leached nitrate is derived from the vast reserves of organic nitrogen in the soil, rather than from fertilisers.

The main factors which influence the extent of nitrate leaching include land use, soil characteristics and climate. In terms of land use, nitrate leaching tends to be greatest when there is no crop cover to utilise the nitrate released from fertilisers or organic reserves. With spring planted cereals, for example, the leaching risk is greatest during the wet autumn and winter period when the soil is left bare.

In order to maximise uptake and minimise leaching, fertilisers should be applied just before and during the period of maximum crop growth, and large applications at any one time should be avoided. Tillage practices also have an effect on nitrogen losses through leaching and denitrification.

Tillage improves topsoil drainage and aeration and, as a result, rates of organic matter decomposition increase. Expert compared nitrogen losses by leaching and denitrification on conventionally cultivated and direct drilled land in southern England. Leaching losses were considerably greater from the conventionally cultivated land, whereas denitrification losses were smaller. For both tillage regimes, however, maximum nitrogen losses occurred during the autumn period after the harvest, and during the spring following fertiliser application.

Nitrate leaching is not restricted to arable land. Old clover-rich grasslands in particular are able to fix large quantities of nitrogen, and this is released when the grassland is ploughed up. This problem is not as serious in short-term arable leys where nitrogen fixation is considerably less.

Even land under needleleaf forestry is not immune to the nitrate leaching problem, although rates of nitrogen release are relatively low in the often acidic and waterlogged soils. Leaching occurs most frequently during the land preparation stages, prior to planting, where improved drainage and aeration lead to increased rates of organic mineralisation, and after felling when the source of nitrogen uptake has been removed.

With respect to soil characteristics, coarse-textured soils are particularly susceptible to leaching due to the often poor structural development and relatively large pore sizes. Rapid drainage in such soils often leads to nitrate contamination of ground-water supplies, particularly in areas which are intensively cultivated. In contrast, clay soils are often characterised by good structural development, small pore sizes and restricted drainage.

Under these conditions, surface runoff may occur and leached nitrate is more likely to be transferred into surface waters than into ground waters. Waterlogging and resulting anaerobic conditions, often found in clay soils, may lead to significant nitrogen losses through denitrification. In soils of tropical environments, which often possess significant anion exchange capacity (AEC), nitrate may be retained in the soil to some extent, thus restricting the leaching process, but this situation does not apply in soils of temperate environments where AEC is usually negligible or non-existent.

In terms of climatic influences on nitrate leaching, rates of mineralisation of organic nitrogen increase with increasing temperature and are particularly high in warm, moist conditions. The timing of individual rainfall events, in relation to mineralisation and fertiliser application, is also important. If a significant amount of rain falls within a month or so of a phase of active mineralisation, or fertiliser application, then nitrate leaching is likely to occur, particularly if there is no crop to take up the released nitrogen, or if uptake is restricted due to poor crop growth.

Experts found that levels of nitrate in rivers draining into Lough Neagh, Northern Ireland, were positively associated with nitrogen fertiliser application rates and December to May river flow, and negatively associated with April to September rainfall and June to September air temperature.

Increased levels of nitrate in surface and ground waters have been implicated, together with phosphate, in a number of environmental and health problems. In surface waters, increased nutrient levels lead to excessive algal growth and oxygen depletion, a process known as *eutrophication*. Stagnation, and shading of the bottom environment by algal blooms, causes increased fish mortality and restricted aquatic plant growth.

Eutrophication is not restricted to fresh water, but may also occur in marine environments. In the North Sea, for example, algal blooms have had an adverse effect on salmon and trout farms off the coast of Norway; this damage has incurred costs in excess of $200 million. In spite of this problem, however, countries of the European Union continue to dump 1.5 million tonnes N a^{-1} into adjacent marine areas, more than 60 per cent of which is derived from agricultural runoff. The nitrate in potable water supplies has important implications for human health.

The incidence of 'blue-baby syndrome' (methaemoglobinaemia), for example, is often greatest where nitrate concentrations in water supplies are high. Similarly, the incidence of disorders of the digestive tract, including stomach cancer, are also associated with elevated nitrate levels, although the medical evidence is by no means conclusive.

In response to these environmental and health issues, legislation has been established in an attempt to control levels of nitrate in potable water supplies. The 1980 European Community 'Drinking Water Directive' (80/778), which actually came into force in 1985, set a limit of 50 mg l^{-1} of nitrate for drinking water; this is equivalent to 11.3 mg l^{-1} of $NO_3 - N$ (nitrate nitrogen).

The latest European Union Directive aims to reduce this value by about 50 per cent to 5.7 mg l^{-1} $NO_3 - N$. In Britain, nitrate levels in drinking water are greatest in intensively cultivated areas where water supplies are derived from deep aquifer sources, particularly in the east and south of the country; these sources contribute about 40 per cent of the potable water in England and Wales.

In the USA, the Drinking Water Standard sets the legal threshold for $NO_3 - N$ at 44 mg l^{-1}, although this value is commonly exceeded in the more intensively cultivated areas of the mid-west. In Kansas, for example, 25 per cent of rural drinking water wells exceed this limit.

Management and remediation of the nitrate leaching problem is a complex issue which depends largely on the adoption of sensible landuse strategies. In terms of nitrogen fertiliser application, slow-release varieties should be used in preference to more soluble types. Similarly, the leaching problem may be reduced if $NH_4 - N$ (ammonium nitrogen) compounds are used instead of $NO_3 - N$ compounds, although increased nitrification may lead to soil acidification.

This may be controlled, however, using the specific inhibitor of nitrification, 'N-Serve', in conjunction with the $NH_4 - N$ fertilisers. It may also be controlled using a mixed fertiliser-lime compound such as 'Nitrochalk'. In landuse regimes where the soil is bare for significant periods of time and the risk of nitrate leaching is high, a 'catch' crop may be grown such as white mustard, forage rape or rye-grass.

Such crops take up the excess nitrogen and are then ploughed into the soil just before the next crop is planted; the nitrogen retained by the catch crop then becomes available to the new crop. The creation of riparian buffer zones adjacent to water courses can also contribute to the immobilisation of excess nitrate, thus reducing the risk of surface water contamination.

In a broader landuse context, the 'Nitrate Sensitive Areas' (NSA) and 'Nitrate Advisory Areas' (NAA) schemes, established in central and eastern areas of England where nitrate concentrations in drinking water regularly exceed the European Community 50 mg l^{-1} threshold, provide a useful set of guidelines for management of the nitrate leaching problem.

The NSA scheme relies on the willingness of farmers to improve agricultural practices on a voluntary basis, with a view to reducing the amount of nitrate leached from their land. Guidelines are issued and farmers are encouraged, through incentive payments, to comply with them. The scheme has two levels—basic and premium. The options for these levels, in terms of agricultural practices.

The NAA scheme aims to promote good agricultural practice but is not supported financially. Its methods are to avoid exceeding the recommended optimum application levels of fertiliser for each crop, to avoid nitrogen fertiliser applications in the autumn, to avoid large nitrogen fertiliser applications at any one time, to minimise autumn ploughing of grassland, and to avoid application of nitrogen fertiliser to hedge bottoms, field margins and water courses.

In addition to the environmental problems associated with fertiliser application, a number of problems arise from the use of pesticides. A wide range of pesticides has been developed (more than 450 compounds), the types most commonly used being insecticides, fungicides and herbicides; other varieties include nematicides, miticides, rodenticides and molluscicides. Pesticides behave in a variety of ways following application.

They may be degraded either biologically or photochemically, adsorbed by organic matter, clay and oxides/hydroxides of iron and aluminium, washed into water courses by leaching (especially compounds with solubilities > 10 mg l^{-1} such as simazine, bromacil and aldicarb) and surface runoff, or they may undergo volatilisation into the atmosphere (especially surface applied compounds, and those of low solubility and vapour pressure such as organochlorines).

Ideally, pesticides should control only the target organism and persist for long enough to achieve this before degrading into harmless products. This is not always the case, however, and a number of environmental concerns arise from pesticide use. These include persistence in the environment, toxicity in soil, vegetation and water supplies, and impact beyond the target organism, including bioaccumulation and its implications for human health

In fact, the degradation products of some pesticides may be as toxic as the original source chemical but are not often measured by standard assays. The persistence and toxicity of many pesticide compounds and their degradation products are also dependent on a number of soil characteristics, notably clay and organic content, and pH.

The environmental impacts of pesticide use were felt most strongly with the early generation of pesticides, particularly the organochlorine compounds and those containing heavy metals. Generally, the organochlorine compounds are the most persistent of the pesticides and may survive for a number of years before they are degraded. Their residues have been found widely in soils, freshwater sediments, fish and cows' milk.

Due to their high lipid solubility, they have also been found in the fatty tissues of animals. All of these findings have important implications for the ecology of food chains and thus for human health. An additional problem was that the effectiveness of organochlorine pesticides decreased as target organisms became resistant to them. The more recently developed organophosphate and carbamate pesticides are much less persistent than the organochlorines, with a half life of about six months, but are often more toxic.

The phenoxyacetic acids, 2, 4-D and 2, 4, 5-T, are also rapidly degraded but have been associated with animal and human growth and reproductive abnormalities.

In terms of management and remediation of the environmental problems associated with pesticide use, continued research and monitoring, with the aim of minimising effective persistence and toxicity, and maximising specificity, are essential. Attempts have been made to model the behaviour of pesticides in soils under different management regimes, with a view to reducing the risk of ground and surface water contamination.

In addition, the use of granular slow-release pesticides is being explored, together with ultra-low volume application techniques. The environmental problems may also be alleviated by adopting alternative, non-chemical strategies of pest, disease and weed control. These include direct approaches, biological and cultural methods and habitat removal. Direct approaches are aimed at clearly identified animals and plants, and specific practices include hunting and hand-weeding.

Biological methods involve the use of predatory species, preferably with a wide environmental tolerance range, to control the specific target organism. Ideally, as the pest population grows, the predator population should follow, and vice versa; this is known as a *density-dependence relationship*. If such a relationship does not exist, however, the predator itself may become a pest.

Cultural methods involve the use of tillage, crop rotation, fertiliser application, liming and drainage techniques. These help to disturb the cycle of pests, diseases and weeds, and to improve the resistance of crops to them.

Habitat removal is perhaps the least acceptable approach to pest, disease and weed control. The basis of this approach is that semi-natural vegetation acts as a refuge for many pests, diseases and weeds, and that control may therefore be achieved through the removal of such habitats. However, this approach fails to recognise both the conservation value of these habitats and the fact that they may be an important source of predators as well as pests.

Woodland and hedgerows in lowland arable areas, for example, are an important habitat for ladybirds, essential in the control of greenfly populations, and for bees, important in plant pollination. They also act as shelter belts, thus helping to control soil erosion.

Urban and Industrial Pollution

Urban and industrial development has been associated with both physical degradation and chemical contamination of soils. Problems of physical degradation include erosion, compaction and structural damage resulting from construction activities and opencast mineral extraction. Similarly, chemical problems result from waste disposal activities, discharge and spillage of liquid effluents, and atmospheric emissions, including acid deposition.

Soils of urban and industrial environments are every bit as complex and variable in their characteristics as those in rural areas and merit classification in their own right. Experts suggest that the England and Wales soil classification system should be modified to include an 'anthropogenic' major soil group which could be subdivided on the basis of the types of contaminant materials present.

Physical disturbance and chemical contamination of soils in urban and industrial environments are not only recent phenomena. Archaeological investigations have revealed the extensive accumulation of building and domestic waste, although much of this was relatively harmless. Since the Industrial Revolution of the eighteenth and nineteenth centuries, however, the amount and variety of waste materials have increased dramatically.

Furthermore, much of this waste is considerably more harmful and often less biodegradable than its early historical counterparts. It is particularly important that any survey of contaminated sites should include an assessment of this historical legacy, since the response of soils to stored pollutants may be delayed for long periods of time. This type of delayed response is potentially very serious and is often referred to as the 'chemical time-bomb' effect. Scientist identifies four major sources of soil contamination in urban and industrial environments—construction and demolition waste, metalliferous materials, power generation emissions, and chemical and organic wastes. During construction and demolition a range of materials are disposed of in the soil environment, including broken bricks, tiles, glass, timber, piping, wiring and cables, insulation materials, mortar, concrete and plaster.

Once in the soil, these materials undergo a number of chemical changes. Plaster, for example, contains large amounts of gypsum and at sites where the water table is high, the gypsum is dissolved and capillary action may bring it into contact with new concrete structures, thus leading to serious corrosion problems. Similarly, although the use of asbestos and its removal from existing buildings is now carefully controlled, significant quantities can be present in soils and overburden at sites with a long history of construction and demolition.

Metalliferous wastes, particularly heavy metals (e.g. lead, zinc, cadmium, copper and nickel), are commonly found in soils of areas where ore extraction and smelting have occurred. More locally, metal contamination occurs on land used for scrap metal dealing and munitions factories. In an attempt to provide systematic data on the extent of heavy metal contamination in urban soils in Britain, a national survey was commissioned by the Department of the Environment in 1981; 100 household gardens were sampled in each of 53 cities, towns and villages.

In most of these localities, total lead levels in the soil exceeded 200 mg kg^{-1}, with values in excess of 500 mg kg^{-1} in industrial areas in the south and in former lead mining areas in the north. These values are considerably greater than the background levels of lead (usually <100 mg kg^{-1}) normally found in uncontaminated soils.

Toxic metals may exist in the soil in a number of forms including adsorbed cations, attached to clay and humus colloids, and organo-metallic chelates. Their availability to plants depends on a number of soil characteristics, particularly cation exchange capacity (CEC), pH and the interdependence effects of other metals. In soils with a low CEC, the metals are not retained effectively and are likely to be either leached from the soil or taken up by plants, while in soils with a high CEC, they are likely to be fixed in the soil through adsorption processes.

Similarly, the mobility and availability of heavy metals is considerably greater in acidic soils ($pH < 5.5$) than in near neutral or alkaline soils. Once mobilised, the metals may enter the food chain either through water supplies and aquatic organisms, or through arable produce and grazing animals. The effects of toxic metals in soils on human health are unclear, and it is therefore difficult to establish threshold concentrations above which toxicity problems are likely to occur. However, a number of countries have devised such threshold values, although these vary markedly in response to differences in environmental policy and legislation. The Dutch government, for example, has a system with three levels—an acceptable reference or background value, an indicative value for further investigation and an indicative value for clean-up.

A number of soil contaminants are derived from the power generation industry, including SO_2 (sulphur dioxide) from coal-fired power stations, and radionuclides from nuclear power stations and weapons testing. The anthropogenic radionuclides most commonly found in soils are those of caesium (^{137}Cs and ^{134}Cs).

Much of the research in this field has occurred since the accident at Chernobyl in Ukraine in 1986, which resulted in large quantities of radionuclides being deposited across wide areas of Europe, including Britain. The behaviour of radionuclides in soils depends on a number of soil characteristics, particularly clay content and mineralogy, organic content, CEC, pH, NH_4^+ (ammonium) content and nutrient status. The radionuclides are immobilised most strongly in soils with a high CEC and near neutral pH values, where they are adsorbed onto clays, especially micaceous varieties, and humic materials.

They are least well retained in acidic soils with a low CEC, where they may be available for plant uptake. Such conditions are widespread in upland areas of northern and western Britain where the milk and meat of grazing cattle and sheep were contaminated following the Chernobyl accident.

In northern Scandinavia, reindeer herds grazing on slow growing lichen heath were similarly affected. The problem of radionuclide entry into the food chain from acid soils in Britain was not anticipated because much of the early research here had focused on clay soils with near neutral pH values where radionuclides are relatively immobile.

A wide range of chemical waste materials have been implicated in the contamination of soils. These include derivatives of detergents, fertilisers and pesticides, paints, dyestuffs, battery chemicals and leather tanning agents, to name but a few. Even silicon chip manufacture involves the use of chemical components, and leakage of solvents used in this process led to the contamination of drinking water in Silicon Valley, California.

In recent years, particular concern has arisen regarding the production of dioxins; these are a particularly toxic group of chemicals and are associated with the manufacture of several organic pesticides. In the Italian town of Seveso, for example, an area of 1,800 ha was affected by dioxin contamination in 1976 and 700 people were evacuated; subsequently, the contaminated soil had to be excavated and removed.

A major event which publicised the issue of chemical contamination of soils was the Love Canal disaster near Niagara, USA. Here, extensive dumping of chemicals over a 30-year period led to severe soil contamination by migrating leachates and gasses, with serious health implications for local people.

A number of organic wastes may lead to soil contamination. At Times Beach, Missouri, for example, the land was so badly contaminated by dioxins from waste oils, derived from the manufacture of the defoliant 'agent orange', that it was purchased by the Federal government.

Spillage of oils and related wastes is common on land used for storage and maintenance of motor vehicles, and for fuel storage. Sewage sludge, often applied to agricultural land adjacent to urban and industrial areas, often contains high concentrations of heavy metals. The PCBs (polychlorinated biphenyls) are a group of organic solvents commonly implicated in soil contamination. These are used as dielectric fluids in transformers and are often released during the break up of electrical equipment.

Management of contaminated soils is a difficult prospect as the guidelines on threshold toxicity levels of contaminant materials, and on the vulnerability of soils to pollution, are still being developed and refined. Management options depend to a large extent on the geological and hydrological characteristics, and size, of the contaminated area.

Such areas vary dramatically from large designated landfill sites with records of waste disposal activities, to former industrial sites with a long and unclear history of pollution. Management options also depend on the nature of the contamination problem and whether it can be contained at the affected site, or whether leachates and gases threaten areas beyond it. Furthermore, it may be possible to treat the contamination on site, or it may be necessary to remove and relocate the contaminated soil.

Experts outline a number of management approaches to the problem of contaminated soil. These include the use of physical stabilisation, barrier, thermal and microbiological techniques. Stabilisation techniques involve treatment to reduce the solubility and mobility of the waste materials.

This is achieved through the application of cement, lime, gypsum, silicate materials, epoxy-resins, polyesters or asphalt; these act as binding agents and help to stabilise the 'landform' within which the contaminated waste is stored. Barrier systems usually rely on physical containment using steel or concrete piling to prevent downward and lateral migration of toxic materials. Similarly, layered cover systems are often used to prevent upward migration of contaminants.

Thermal techniques involve heating contaminated soil in rotary kilns or furnaces, whereas microbiological techniques involve the inoculation of waste materials with microbiological communities. Both of these techniques aim to convert the toxic waste materials into less harmful forms.

Other approaches to the management of contaminated soil include chemical treatments which may be used to hydrolyse or oxidise contaminants into less dangerous products; similarly, acidic or alkaline wastes may be neutralised. In addition, physical methods may be employed to separate out contaminants according to particle size or density.

The number of sites officially designated as contaminated varies dramatically from country to country. In the UK there are 300 sites, covering an area of about 10,000 ha, where land is officially designated as contaminated. Unofficially, however, there may be between 50,000 and 100,000 sites, covering an area of over 100,000 ha.

In the USA 25,000 sites have been designated as contaminated by the Environmental Pollution Agency. Similarly, in the Netherlands 6,000 sites are currently being reclaimed. Experts estimated that the cost of dealing with contaminated soils amounts to about US $8-12 per head in the industrialised countries.

This Figure does not include the former Eastern-bloc countries, however, as it is only in recent years that the rest of the world has become aware of the extent and severity of soil contamination here. The problem of contaminated soils is often far too great to be dealt with by individual companies, hence governments are having to provide sizeable financial contributions towards the management of this problem.

Index